11/17

Ricky's Atlas

Mapping a Land on Fire

Ricky's Atlas

Mapping a Land on Fire

Judith L. Li

Illustrations by M. L. Herring

Oregon State University Press
Corvallis

A generous gift from the John and Shirley Byrne Fund for Books on Nature and the Environment helped make publication of this book possible. The Oregon State University Press is grateful for this support.

Library of Congress Cataloging-in-Publication Data

Names: Li, Judy L., author. | Herring, Margaret J., illustrator.
Title: Ricky's atlas : mapping a land on fire / Judith L. Li, M.L. Herring.
Description: Corvallis, OR : Oregon State University Press, 2016. | Sequel to: Ellie's log. | Summary: "Ricky Zamora brings his love of map-making and his boundless curiosity to the arid landscapes east of the Cascades Mountains. He arrives during a wild thunderstorm, and watches his family and their neighbors scramble to deal with a wildfire sparked by lightning. Joined by his friend Ellie, he sees how plants, animals, and people adjust to life with wildfires" —Provided by publisher. | Includes bibliographical references.
Identifiers: LCCN 2015039873 | ISBN 9780870718427 (paperback)
Subjects: | CYAC: Nature study—Fiction. | Fire ecology—Fiction. | Wildfires—Fiction. | Fires—Fiction. | Maps—Fiction. | Oregon—Fiction. | BISAC: JUVENILE FICTION / Nature & the Natural World / Environment. | SCIENCE / Environmental Science. | EDUCATION / Elementary.
Classification: LCC PZ7.L591174 Ri 2016 | DDC [Fic]—dc23
LC record available at http://lccn.loc.gov/2015039873

♾ This paper meets the requirements of ANSI/NISO Z39.48-1992 (Permanence of Paper).

First published in 2016 by Oregon State University Press
Second printing 2017
Printed in China

Production Date: 5.11.17 Batch Number: 78724-0 / EPC 808901.1
Plant Location: Everbest Printing Co. Ltd. Nansha, China

Oregon State University Press
121 The Valley Library
Corvallis, OR 97331-4501
541-737-3166 • fax 541-737-3170
osupress.oregonstate.edu

Dedicated to our husbands

Hiram, who introduced the grand landscapes
and streams of eastern Oregon to Judy

and

John, who was there with Peg when
lightning struck at Waldo Lake

Table of Contents

CHAPTER 1

The Lightning Storm

Ricky Zamora watched bundles of tumbleweed blow like giant beach balls, bouncing across the highway. High, billowing storm clouds raced across the sky. Along the edge of hills to the south, wisps of dust devils spun like mini-tornadoes. Ricky and his mom were on their way east from the Cascade Mountains to his Uncle John's ranch in eastern Oregon. They were caught in a very blustery storm.

A remarkable pair of jet-black ravens fought against the churning wind. They tried to fly forward, but they were struggling, pushed backward with every gust. When Ricky felt the wind make the car sway on the road, he sat up a bit straighter. The storm was all around them. "Hope we beat the rain to Uncle John's ranch," he said.

In his notebook, Ricky recorded wildlife sightings, passing mile markers, and the chaos stirred up by the storm. As they turned up the dirt driveway to his aunt and uncle's house, the wind was still whipping. All kinds of stuff was blowing around the yard—lawn chairs, buckets, stray newspapers flying in the air. "Crazy," Ricky thought. His eight-year-old cousin Sarah managed to wave as she ran past him, chasing a tumbling watering can. With a mighty effort, Ricky pushed open the car door. He and his mom grabbed scraps of papers flying by as they climbed

the porch steps. Aunt Rosa put down the jumble of debris she'd collected to greet them. "Buenos días," she said with open arms.

Sarah stumbled up the steps after them, carrying the watering can she'd saved from the wind. Ricky spotted his older cousin, Tony, scrambling in the garden to check gate latches that would keep chickens in the yard. Max, the family's Australian cattle dog, followed close behind. Farther down the hill Uncle John sprinted for the old wooden barn and pulled the big doors shut. As he and Tony started toward the house, the storm arrived in full force with pounding, driving rain and rolling thunder.

The family, and one wet pooch, scrambled to get inside. They huddled in a dizzy whirl of wet jackets and hugs. "Hola! Welcome! Thanks for bringing rain from the west side," Ricky's uncle said with a wide grin. Tony stomped his boots and shook rainwater off his cap. His dog gave a wild, wet shake, too. Still dripping a bit, Tony greeted his cousin. "Looks like you got here just in time, Rick."

Thunder rattled and rolled again. They counted the seconds between the flash of lightning and the thunder—one one hundred, two one hundred, three one hundred—all the way to eight. That meant lightning strikes weren't too far away. Ricky laughed. "Guess you could say we're starting off with a bang!"

Pronghorn antelope are among the fastest animals on earth. Both males and females have backward-curving antlers and white patches on their rumps.

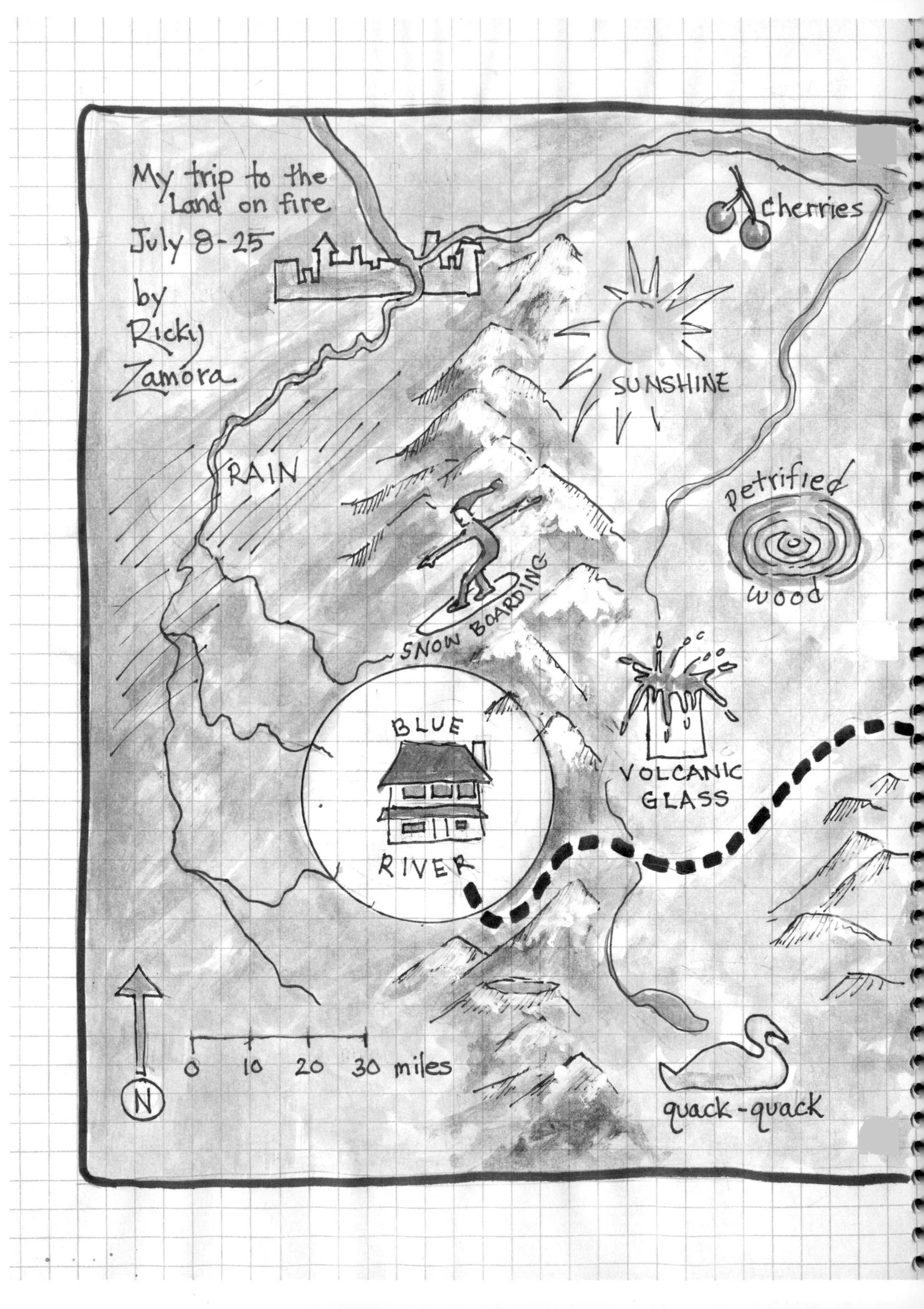
My trip to the Land on fire
July 8-25
by Ricky Zamora
cherries
SUNSHINE
RAIN
petrified wood
SNOW BOARDING
BLUE RIVER
VOLCANIC GLASS
0 10 20 30 miles
N
quack-quack

Prairie
wind turbines
WAGON TRAIL
FOSSILS
ZAMORA
RANCH
FIRE
BASIN
and
RANGE
HOT
SPRINGS
thunder
eggs

Tony chuckled. "C'mon Rick, you're staying with me." He picked up Ricky's backpack while Ricky grabbed his duffel bag and followed his cousin toward the bedroom. Tony was five years older than Ricky, and they always had a good time together. As Ricky threw his duffel on the spare bed, he looked around Tony's bedroom. There were familiar photos of Tony riding his horse and playing baseball. An old lasso hung on the wall next to new blue ribbons from the county fair. Max plopped onto his doggie pillow next to Tony's bed while Ricky stashed his clothes in a drawer. Boots, baseball cap, and backpack were stored in the closet, his notebook on the nightstand.

Cattle dogs, like Max, were first developed in Australia to help people herd cattle over long distances.

Back in the kitchen Aunt Rosa had dinner set at the big, long table. The family ate and chattered while the storm crackled around them. Finally, as Ricky and his cousins finished dinner, the noisy part of the storm moved on. After clearing the dishes, they stepped out on the front porch and into the evening's cool air.

"Wow, look how the wind blew away the clouds," said Ricky. The bright moon was rising in the clear, darkening sky, and stars were appearing like tiny twinkling lights.

"That was some kind of noisy thunderstorm, but the rain gauge says only one inch," Tony observed. He picked up the little notebook they kept on the porch railing and jotted down the record.

"No way," Ricky said. "That's like the gauge we have at home in Blue River."

"Our records go into a national website. How about you?"

"We do that, too. We could look at the differences between your station and ours on their daily maps. I've gotta tell my friend Ellie."

Meanwhile Sarah took a deep breath, just enjoying the moment. "Mm, everything smells so fresh," she said with a little sigh.

Ricky copied her, taking a deep breath of the rain-washed air. It didn't smell like the forest back home after a rain. That was a wet, earthy, mushroomy smell. This was tangy, spicy. It'd been more than a year since he'd been on the ranch, and the familiar fragrances were like a welcome-home greeting. "For sure, I'm back on the east side," he declared.

For a delicious end to a long day Ricky and his mom brought out a big bowl of ripe blackberries they'd picked in Blue River before they left home. Aunt Rosa fetched bowls, spoons, and ice cream for everyone. She loved the wild blackberries that seemed to grow everywhere on the west side of the mountains. "We just can't grow berries like these on the east side," she said as she added another spoonful of berries to Sarah's bowl. "Just picked yesterday," Ricky replied with a berry-stained grin.

Big sagebrush can survive in very dry places because they have taproots that can reach 12 feet down. Their leaves smell spicy after rain.

Ricky's day had begun in the early morning, and he was tired from the long car ride. Before he shut off the bedroom light, he started sketches of a funny idea he had for a map of his trip. Cartoons of what he saw, mile markers—a project to finish later . . .

The next morning, there were chores to be done around the ranch after the wild storm. The Zamora family raised alfalfa that fed livestock on several nearby ranches. "Tony, let's start

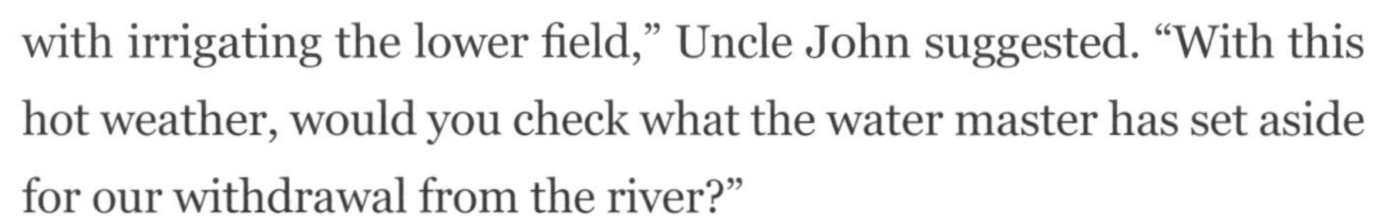

with irrigating the lower field," Uncle John suggested. "With this hot weather, would you check what the water master has set aside for our withdrawal from the river?"

"Maybe after that rain he'll have new calculations for us. I'll look online—meet you down at the pump," Tony called to his dad, who was already out the door. A few minutes later Tony was on his way to help open the water gates.

From the front porch, Ricky watched the sprinklers spring to life in the alfalfa fields. He knew his big cousin had important jobs around the ranch. When Tony started walking toward the old barn, Ricky decided to join him. He skipped down the porch steps, then followed the dirt path that bordered the deep green leaves and purple blooms of alfalfa.

"Over here, Rick," Tony called from the barn. Max barked a friendly greeting and trotted uphill. Ricky gave him a rub behind his ears before running toward the barn. He and Max stepped through the wide-open door into the barn's cool, damp air. Tony was cleaning the horse stall. "Barney, meet my cousin Ricky. Barney's my working horse," he explained. "Hold on, I'm just finishing up with sweeping right now."

While Tony finished his chores, Ricky surveyed the wall where old and newer saddles, ropes of many lengths, bridles, and horse equipment of all kinds were hung. The big barn was built from really big timbers. Bending way back, Ricky could see huge, high rafters. "Those must've been some big old trees," he said.

Tony nodded as he put away his tools and wiped his hands on his jeans. "Yep, they probably came from ponderosa pine forests up the hill, maybe a hundred years ago.

Tony gave Barney a little pat. "Gotta go now, big boy. Time to check the water levels in the field." The sun was high in the sky as he and Ricky walked to the water gate. With a quick twist of the knob Tony shut the flow of water for the day.

Ricky began to feel sweaty hot. "How hot do you think it'll get today?"

"Hmm. Seems like an ordinary kind of hot day for us. I'd guess it'll peak around ninety-eight. Let's check the temperature gauge at the house."

The gauge that hung on a beam of the front porch read 90°F. "Pretty good guess, Tony."

"It'll get hotter for sure—before yesterday's storm the temperature was over a hundred degrees. The air's dry, so it's not so bad, especially in the shade. Just the same, a swim in the river would feel good, don't you think?"

"Alright!" Ricky declared enthusiastically.

So after lunch Ricky, Tony, and Sarah went for a dip at the Zamoras' favorite swimming hole. When they got to the river, Tony opened his backpack. He'd brought a little extra gear for his cousin. "Here's a mask and snorkel so you can see underwater, Rick."

They splashed around in a deep pool to cool off, then waded toward a shallower section. Where water spilled over the smooth rocks like a miniature waterfall, Ricky found a perfect place to watch fish. Tony showed him how to hold himself in place by pressing onto the rocks with both hands. "Like a real live aquarium," Ricky thought, looking through his mask. A big rainbow trout was at the head of the pool, ready to catch insects spilling in. Smaller, less aggressive fish were hanging around the edges. Suddenly Ricky

You can get a rain gauge and become part of a national network of record keepers with the Collaborative Community Rain, Hail, and Snow Network. (www.cocorahs.org)

popped out of the water, giggling. "Those little speckled dace are tickling my feet." He bent over to watch them. "They're nipping on the stream bottom. Searching for food, I'd guess."

Meanwhile Sarah was gliding quietly just downstream, tracking the stream bottom with her mask. Ricky slipped back into the water; he was letting the stream gently float him closer to his cousin when Sarah pointed toward the sand bar they were passing. He turned his head to get an underwater look at several, maybe twenty, dark shells pointing out of the sand. He pulled one out to discover a blue-black freshwater mussel, about two inches long. Instantly it "clammed up"—both valves shut tight. "Oops," he thought, returning it gently to its sandy home.

They floated farther downstream until they found a sloping bank where they got out. While they were walking upstream, Sarah stopped abruptly. "Sh," she said. A small, furry head popped out of the grassy bank, then slid into the river.

It was a slender, tan critter with a tail almost as long as its body. "Mink," Ricky decided.

"Probably looking for its burrow," Tony said. He pointed to the bank on the other side of the river. "That little hole looks about right for the skinny guy," he observed just as the mink emerged from the water, gave a quick shake, and slipped into his burrow.

"Nice," Ricky whispered. He was settling into the rhythms of the east side. The dry, hot weather, a cooling swim in the river, revisiting favorite spots—life here was pretty great.

After dinner an evening quiet settled over the valley. Feeling full and happy, Ricky stepped onto the porch. The juniper and sagebrush smells of the high desert greeted him. Tony and Uncle

John were already there, looking toward the horizon. A few ridges to the south, the air was beginning to glow a weird orange. Ricky squinted into the evening twilight. "Forest fire?"

Tony nodded. He pointed toward smoky wisps that seemed to grab the twilight sky like long dark fingers. "That's a forest fire."

Date: July 8 recorded by R. Zamora

Location: heading east toward Zamora Ranch

Weather: 59° (when we left Blue River)

99° (when we got to the ranch)

Cumulonimbus thunder clouds

Notes: I'm going to make an ATLAS about the eastside of the mountains

Where to start??

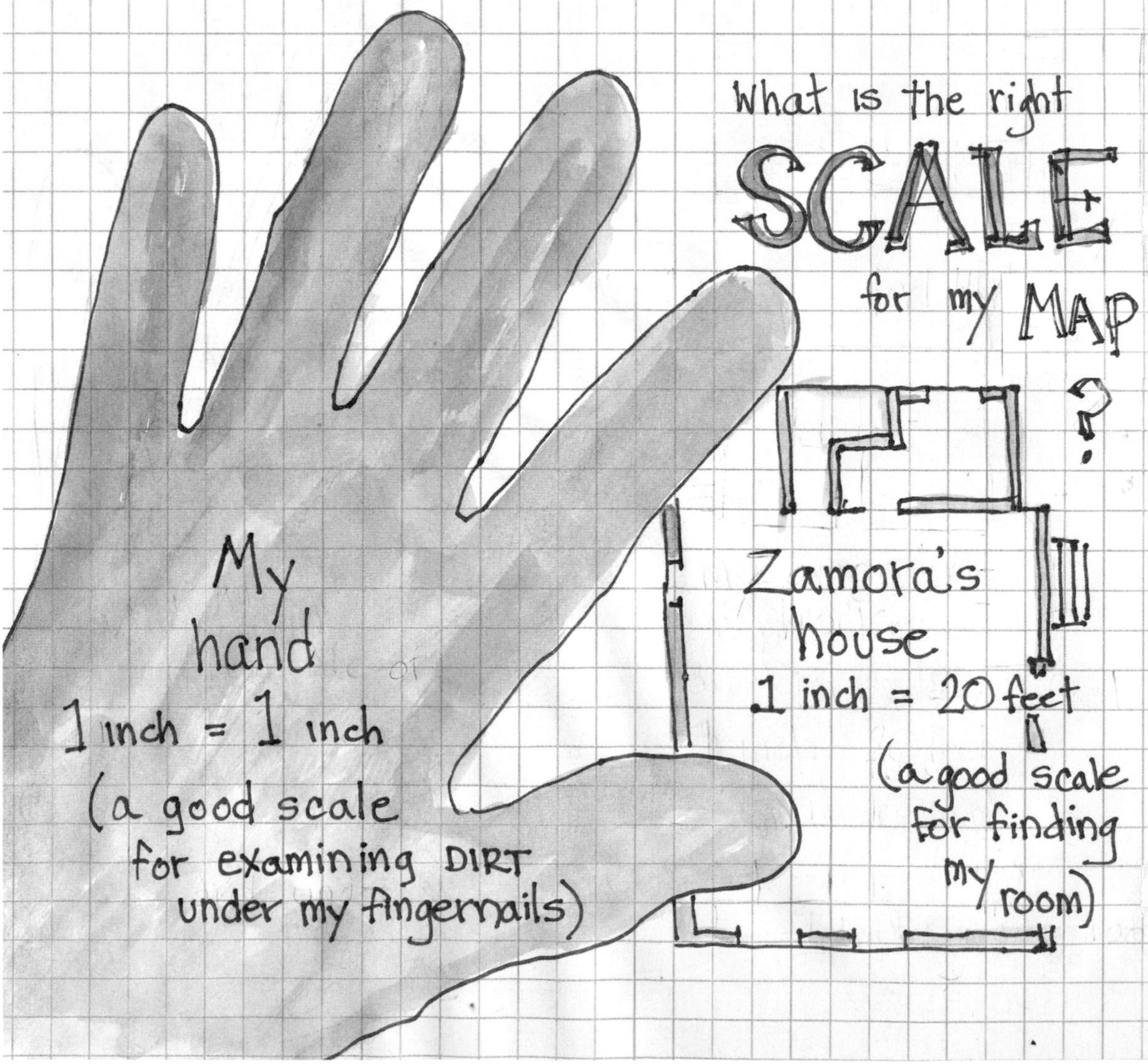

What is the right

SCALE

for my MAP?

My hand

1 inch = 1 inch

(a good scale for examining DIRT under my fingernails)

Zamora's house

1 inch = 20 feet

(a good scale for finding my room)

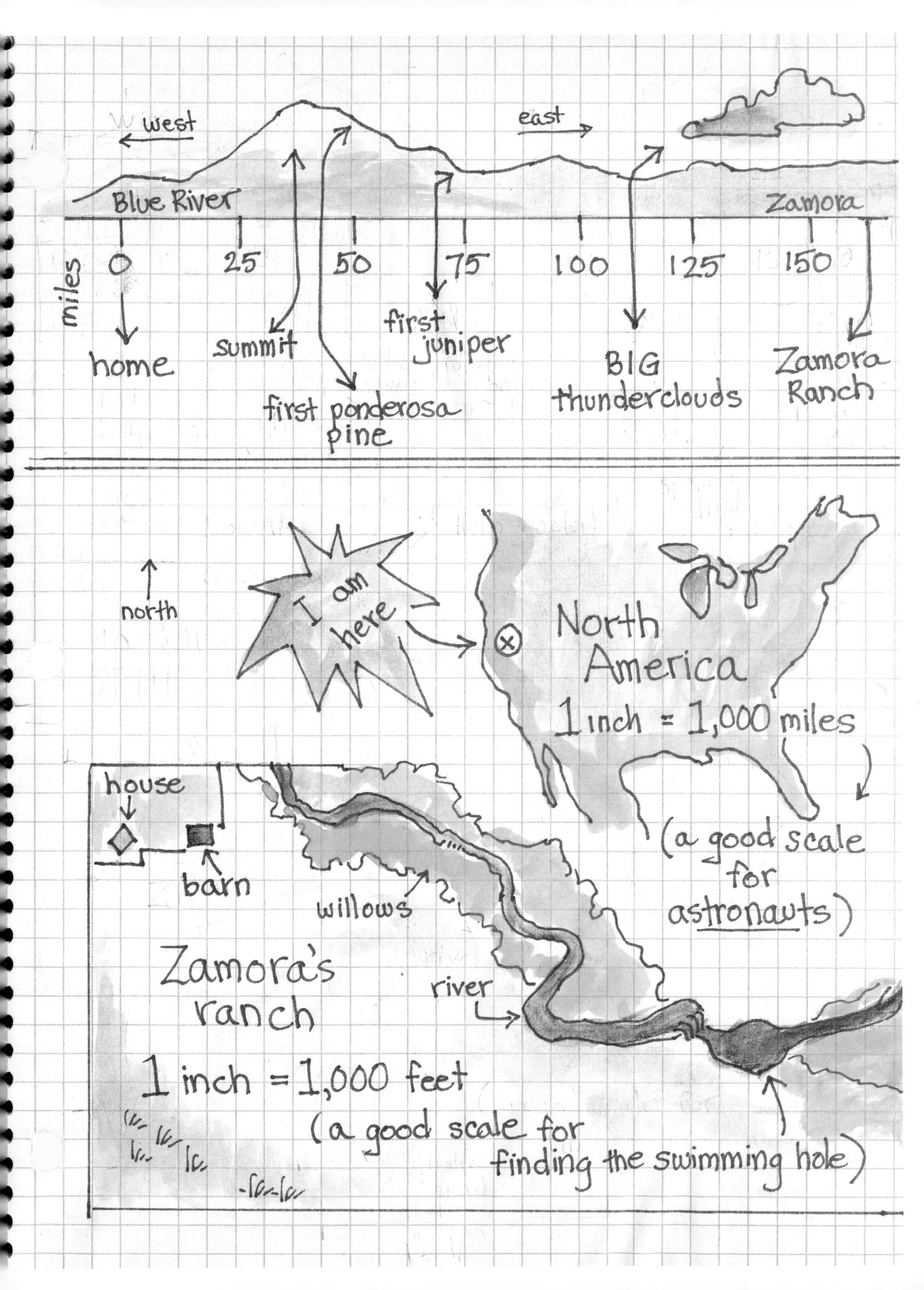

west
east
Blue River
Zamora
miles
0
25
50
75
100
125
150
home
summit
first ponderosa pine
first juniper
BIG thunderclouds
Zamora Ranch
north
I am here
North America
1 inch = 1,000 miles
(a good scale for astronauts)
house
barn
willows
Zamora's ranch
river
1 inch = 1,000 feet
(a good scale for finding the swimming hole)

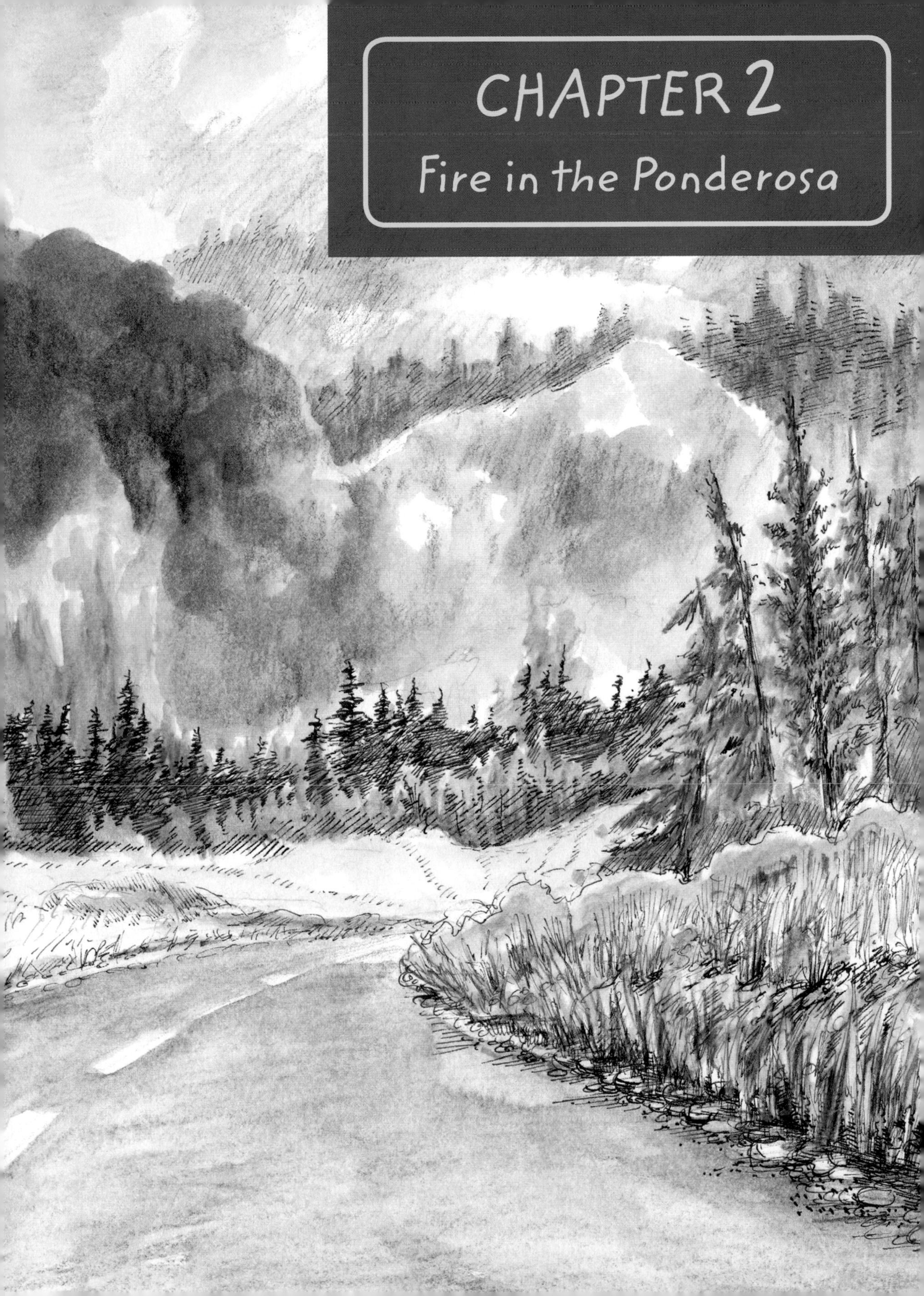

CHAPTER 2
Fire in the Ponderosa

Thup-thup, thup-thup. Ricky ran to the far side of the porch. "What's that?"

"Helicopters overhead," Tony reported. "They're carrying water to the fire site."

"Impressive buckets," said Ricky, watching silhouettes of the huge vessels being whisked through the darkening sky.

Forest Service fire trucks, lights blinking, sped down the road past the ranch. "Fire isn't far away," Tony declared.

Ricky saw the worry on his mom's face. "Whatcha thinking, Mom?"

"Forest fires are more common here than on the west side," she said, remembering the years when they lived nearby, before they moved to Blue River. "It's a lot drier here. I'd forgotten how quickly a lightning strike can flare into a forest fire."

This was the first time Ricky could remember being so close to a fire. Exciting, and scary. That night the family was on high alert, keeping watch for signs of fire on the closest ridgeline. "Can we tell where the fire will go?" Ricky asked.

"Winds from yesterday's storm are gone—it's hard to tell which way the fire might burn," Uncle John explained. Everyone took turns between fire lookout duty from the porch and gathering important things they might need if they had to evacuate. Ricky and Sarah made sure they had a few of their own clothes, and they stayed up watching with their folks late into the night.

Early the next morning, Aunt Rosa checked the fire website for the latest news. "¡Dios mío!" she exclaimed. "They say the fire was started by a lightning strike during the thunderstorm south of us, near Coyote Creek. There are already ten firefighting

units working on it. It's about 25 percent contained, mostly at the eastern edge."

She looked up from the computer. "Fire camp is on River Road. John went up there before sunrise to help set up communication lines."

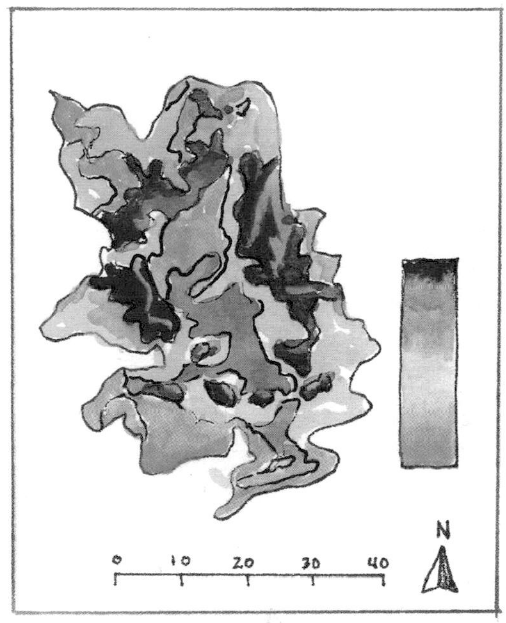

Ricky's mom looked worried. "Maybe we'll get a cooler, non-windy kind of day to slow it down," she said.

"Dumping water on the fire won't do the trick?" Ricky asked.

"Right. When a wildfire gets going, calm air and cooler temperatures help the firefighters a lot. That's hard to predict in the middle of summer," his mom told him.

Tony had been reading the fire report over his mother's shoulder. Quickly, he grabbed his cap and bounded for the door. "I'm off to Henry Hixon's to see if he needs any help," he shouted, grabbing his phone on the way out. Henry was the old cowboy who owned the ranch next to theirs, and Ricky had heard lots of colorful stories about him. He figured Tony would take the quickest route to Henry's—riding his horse straight to the ranch.

Maps online continually update wildfire activity.

As Tony shut the front door, Uncle John called home with important news. "They think the fire began when lightning struck a juniper tree—burst into flame and ignited the grasses. The fire got blown uphill and really got going when it reached the ponderosa pine forest. That's what's burning now."

"Where's it headed?" Aunt Rosa asked urgently.

Helicopters can carry buckets with up to 400 gallons of water and fire retardant. When the helicopter is in position, the crew releases the water to put out hot spots.

"It seems to be moving several miles south of our place," he told her. "But keep track of the bulletins." A snapping crackle interrupted the call, then Uncle John came back on. "Could you bring me a change of clothes? Setting up has been a really gritty business. I'm going to be here quite awhile."

"Of course. Tony's on his way over to check on the Hixon ranch. We'll wait to hear from him before we take off."

"Good idea. Hope things are okay over there. No rush for me. Gotta go."

Aunt Rosa turned to Ricky and Sarah. "Would you like to see the firefighters' camp?"

"Would I ever!" Ricky hurried to grab his baseball cap; Sarah found her camera. By the time Aunt Rosa had collected what she thought Uncle John might need, her phone rang again. It was Tony. "Henry Hixon's cows are grazing up in the forest. They've got to be rounded up away from fire danger. I'm off to help Henry find 'em."

"Is anyone else going with you?" his mom asked.

"Yep—his regular crew and some of the other neighbors. It'll take most of the day I'd guess."

"Okay. Let us know if you need anything." Aunt Rosa told him. "And be careful!"

"Sure bet, Mom. You know how Henry is. He's seen these fires before; he'll watch out for us." Aunt Rosa smiled and shook her head a little as she put her phone away.

"There are cows in the forest?" Ricky asked.

Aunt Rosa nodded. "Cows are herded up there this time of year. They graze on shrubs in the forest and in meadows higher up. Meanwhile, Henry can grow hay on his lower pastures in the valley. Now, are we ready to go?"

"Absolutely!" Ricky replied, halfway out the door.

Driving down the ranch's dirt road, they spotted a green Forest Service fire truck speeding east toward the fire. By the time they'd turned onto River Road, the truck was already out of view.

"Where's the fire camp, Mom?" Sarah asked.

"Your dad said it's just past mile marker 22—can you two keep a look out for it?"

"Yep—we're on it," Ricky said.

The air was distinctly smoky. As they passed mile marker 20, the road began to climb; it hugged the hillside, then turned sharply south. Off in the distance, in full view: The Fire. Bright orange flames and grey plumes of smoke rose above the black outline of forest trees.

"Wow!" Ricky exclaimed, pressing his face to the window to see the top of the towering, billowing smoke clouds. Sarah pulled out her camera and snapped several photos. Aunt Rosa kept her eyes on the road, driving straight ahead. The road took another turn, and they lost sight of the fire. Ricky and Sarah sat back, speechless, overwhelmed by what they'd seen. No one noticed when they passed the mile marker 21 sign, but pickup trucks and cars parked against the roadside fences were a clue that they were coming up to mile 22. A hand-scribbled sign pointed uphill toward "Fire Camp."

Aunt Rosa turned up the dirt road. A woman in a bright green safety vest signaled them to stop. "This road's closed. Fire camp up ahead," she said.

Hand crews work to stop wildfire from spreading with tools like Pulaskis and shovels that eliminate hot spots.

Smoke from wildfire can travel for hundreds of miles.

"We're delivering supplies to John Zamora—he's helping set up the camp," Aunt Rosa told her.

"Okay, the camp's about a half-mile up. But don't stay too long—there's a lot of coming and going up there." She signaled Aunt Rosa to continue.

"Do you think we'll get any closer to the fire?" Ricky asked.

"Sorry, Ricky—they keep the fire camp a safe distance from the fire to protect the crews," Aunt Rosa replied.

"Right," he answered, just a little disappointed. "Makes sense to keep the firefighters safe." Their car moved slowly up the road, leaving a dusty cloud behind. They drove over a small knoll, and then they saw the fire camp stretched across the meadow below.

"Whoa, look at all those tents."

Sarah eyeballed clusters of ten tents at a time and did a quick count. "There's twenty, thirty, forty . . . geesh, I'd guess more than fifty little tents. And what are those big ones over on the side?"

"That's probably where your dad is helping with the wiring," Aunt Rosa said, driving toward the largest tent.

Ricky and Sarah watched a small group of firefighters walking slowly into camp, outfitted in bright yellow jackets and pants, sturdy helmets, and heavy boots. They were covered in layers of dirt and black soot. They looked tired. “Just think, that crew was somewhere close to that big blaze we were watching from the road,” Ricky said quietly. Other men and women were going back and forth from the cafeteria tent; some seemed in a hurry, maybe getting ready to go out to the fire again.

Aunt Rosa parked the van to the side of the big tent. Out of the confusion Uncle John came walking toward the car. “Welcome to firefighting central.”

“This looks crazy-complicated,” Ricky said, hanging out the car window to get a better look at the action. He was studying the maze of wires and hoses running through the brush, crisscrossing every which way.

His uncle nodded. “It takes a lot of people and a lot of equipment to fight the fire,” he said. A fleet of trucks, tankers, and bulldozers were parked on the edge of the camp. “Still, we’ll probably depend a lot on favorable weather to help contain this one.”

Aunt Rosa handed Uncle John his change of clothes and a sack lunch. “Thanks for bringing all this,” he said. A caravan of three vans, filled with firefighters in full gear, followed by two big tanker trucks, headed up the hill. “Those crews and water are headed toward the fire,” Uncle John said, pointing at the ridge, “and away from our place. Everything okay at home?”

“Sí, yes, everything’s fine,” Aunt Rosa replied. “Tony’s helping Henry bring the cows off the forest. Will you be home tonight?”

"Hard to tell," he said. "I'll give you a call. Keep checking the fire reports."

She nodded. "Will do. Take care." They waved as Uncle John watched them drive down the dusty road, away from the makeshift tent city.

Sarah and Ricky looked for the open space in the road where they'd seen the fire. They watched the blaze as long as they could, eventually turning to look backward until the hills and the road's twists and turns obscured their line of sight. Just as they settled back into their seats, the car came to a sudden, abrupt stop.

A herd of about forty cows was moseying slowly across the road, mooing in a funny cow-chorus. Right behind them Tony, Henry Hixon, and other cowboys on horses nudged the cattle along. Max and three more herding dogs circled around the edges. Tony rode farthest back, shouting at the herd, keeping them on the move, doing his best to speed them up. The cows seemed to be in no hurry, and they stirred up a cloud of dust that lingered after they'd passed by. Sarah and Ricky watched Tony reining in his horse to chase a young calf that had separated from the herd. "Barney responds really well to Tony's signals," Ricky noticed.

Sarah nodded. "They're used to working together." The family waited while the whole procession crossed slowly from the hillside down to the meadow. When Tony finally closed the gate to the field, he wore a thick layer of dust from the top of his hat to the tip of his boots. With a broad grin he waved his family past.

When Tony returned from the Hixon ranch, it was almost dark. The cousins sat around the kitchen table exchanging stories about the extraordinary day. "Henry's crew moves his cattle around a lot, so they knew where the cows were grazing," Tony explained. "That's how we found his cows in the forest today. Took only about half an hour to reach them, but it felt like forever getting them to safer ground. It must've been pretty exciting to see the fire," he told Ricky. "But I was sure glad it wasn't too close to Henry's cows."

Real cowboys. Real hard work. Ricky was fascinated. "Think I could meet Henry sometime?"

Tony grinned. "Sure thing. We'll go over to his ranch while you're here."

That night, Ricky lay in bed smelling a hint of smoke in the air. He could hardly believe so much had happened in the short time he'd been at the ranch. He fell asleep thinking about the fire, the crews rushing to the fire, the cows trailing across the road, and the fire, again.

Black-billed magpies are often seen along country roads. They eat fruit, grains, beetles—even maggots they find in carrion. Sometimes they pick ticks off the backs of cows.

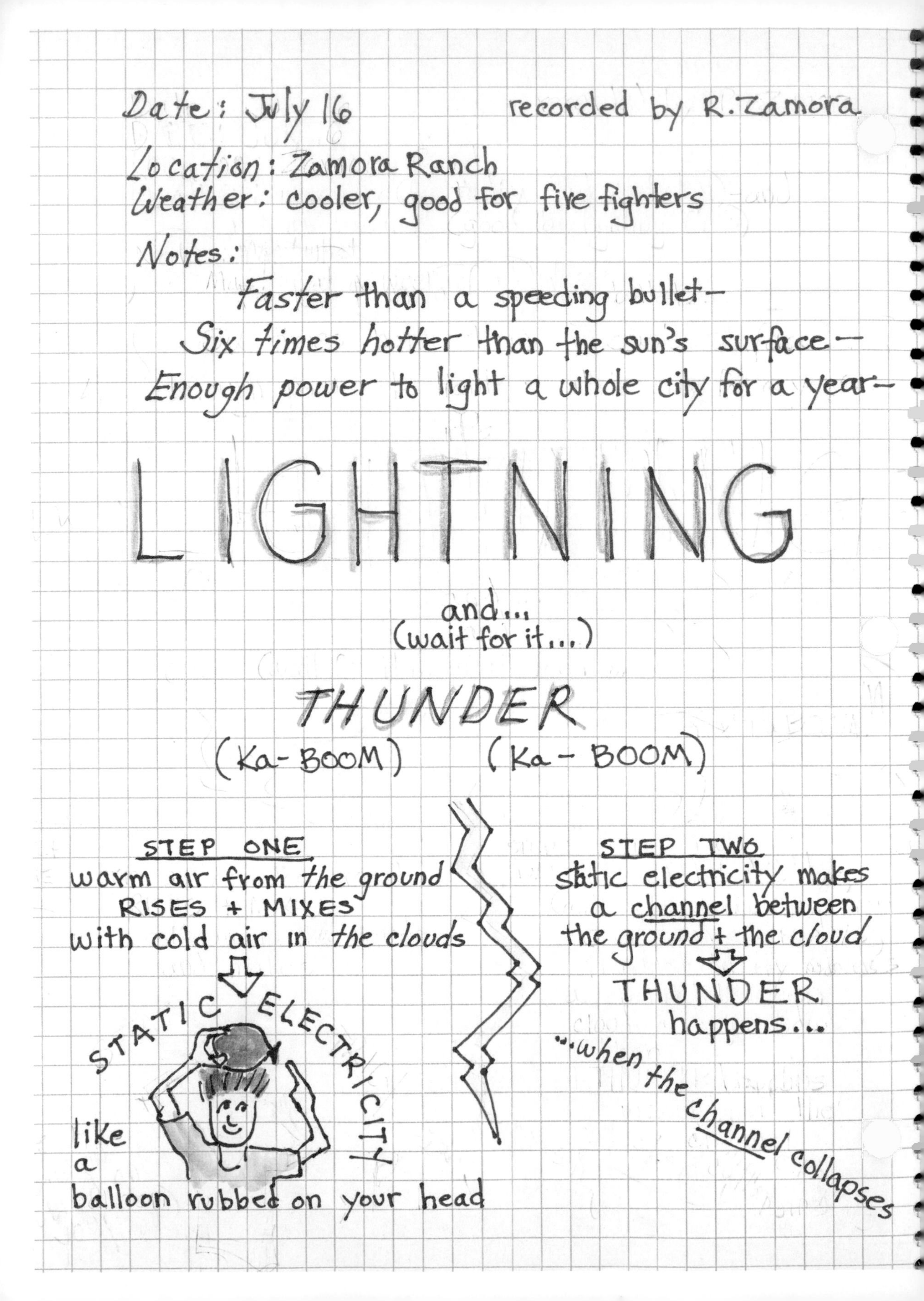

Date: July 16
recorded by R. Zamora
Location: Zamora Ranch
Weather: cooler, good for fire fighters
Notes:
Faster than a speeding bullet—
Six times hotter than the sun's surface—
Enough power to light a whole city for a year—
LIGHTNING
and...
(wait for it...)
THUNDER
(Ka-BOOM) (Ka-BOOM)
STEP ONE
warm air from the ground RISES + MIXES with cold air in the clouds
STATIC ELECTRICITY
like a balloon rubbed on your head
STEP TWO
static electricity makes a channel between the ground + the cloud
THUNDER happens...
...when the channel collapses

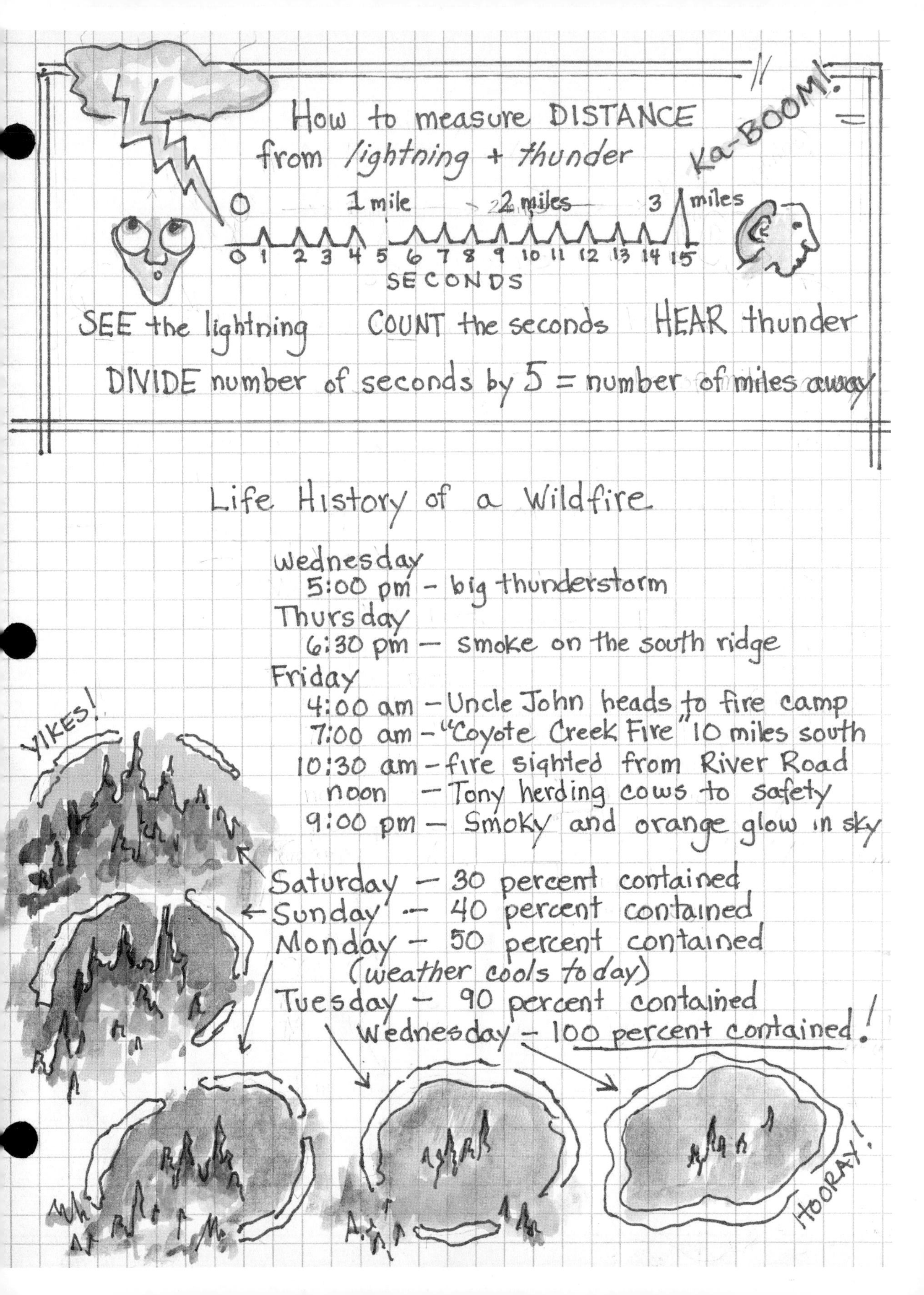

How to measure DISTANCE from lightning + thunder
KA-BOOM!
0
1 mile
2 miles
3 miles
0 1 2 3 4 5 6 7 8 9 10 11 12 13 14 15
SECONDS
SEE the lightning
COUNT the seconds
HEAR thunder
DIVIDE number of seconds by 5 = number of miles away
Life History of a Wildfire
Wednesday
5:00 pm - big thunderstorm
Thursday
6:30 pm — smoke on the south ridge
Friday
4:00 am - Uncle John heads to fire camp
7:00 am - "Coyote Creek Fire" 10 miles south
10:30 am - fire sighted from River Road
noon - Tony herding cows to safety
9:00 pm — Smoky and orange glow in sky
YIKES!
Saturday — 30 percent contained
Sunday — 40 percent contained
Monday — 50 percent contained
(weather cools today)
Tuesday — 90 percent contained
Wednesday - 100 percent contained!
HOORAY!

CHAPTER 3
Eagle's Roost Tower

Early the next day Tony gave a fire update. "The website says 'Calm weather, no winds. The Coyote Creek Fire is about 40 percent contained.' Sure hope today will be more normal."

"Ricky, your friend Ellie's dad called last night," Aunt Rosa announced. "He's coming this way to help with forest surveys and asked if Ellie could stay with us."

"No kidding?" Ricky exclaimed. "That'll be fun. You'll like Ellie, Sarah. There's lots we can show her."

"This is turning into a pretty exciting summer," Sarah said.

"Right now's a good time for you to gather eggs from the chickens," her mother suggested. "We kinda got off schedule yesterday."

Sarah slowly pushed her chair away from the table. "Okay, Mom." She looked at Ricky. "Can you give me a hand?"

"No problem," he said. Sarah handed him a basket and happily led the way to the well-beaten path that crossed the front lawn, past the apple and pear trees, to the chicken coop. Songbirds flitted between the fruit trees, singing their morning chorus, full of chirping, buzzing, and melody. "Dad says we're at the end of the nesting season," Sarah said. "Their best songs are earlier in the year, but this is still nice." One bird with a bright yellow chest perched on the barbed-wire fence and burst into glorious melody. "I know that fellow," Sarah said. "A meadowlark. He likes open meadows—get it, *meadow-lark*?"

In the early twentieth century, the US Forest Service began building fire lookout towers. Most had an Osborne Fire Finder (a rotating steel disc and sighting fork) to accurately pinpoint forest fires by sighting distant smoke.

"Got it," Ricky smiled. He took a closer look at the cheery bird. "Check out the black V across his yellow chest. Makes him look like he's wearing a snazzy vest."

When they reached the coop, Ricky counted about a dozen chickens scratching the ground. Sarah unlatched the little gate and walked toward their covered shelter. "Let's peek in here." She reached into the straw. "Egg number one!"

"Wow, there's a lot of different egg colors," Ricky noticed, picking up dark brown eggs, light brown, green, and even a few white ones.

"Yep, different chickens, different eggs," said Sarah. They filled the baskets and made sure the latch on the gate was secure. While they were walking back to the house, a pickup turned into the driveway. As soon as the dark green truck stopped in front of the house, Ellie Homesly jumped out. "Hey, Ricky!" she shouted. "Where's the fire?"

Sarah laughed. "She *is* a ball of fire, isn't she, Ricky?"

The friends hugged. Ricky introduced Ellie and her dad to everyone, and Ellie gave a bucket full of fresh blackberries to Ricky's aunt.

"I got your text," she whispered to Ricky, as Aunt Rosa's eyes lit up. "Muchos gracias!"

Excited to have a new friend, Sarah quickly ushered Ellie toward her bedroom where the girls settled her things. Over lunch, they recounted the excitement of the fire. Ellie was amazed at the contrasts she'd seen on her trip. "It was raining a little in Blue River when we left home this morning," she said. "When we came over the mountains, it felt like the air got drier and definitely warmer."

"That's the rain shadow," Tony said. "We were mapping it out at school last year. The Cascades draw a lot of moisture from the clouds on the west side before they reach us here. That's why on our east side the lands in the shadow of the mountain get air that's loads drier.

Ellie opened her journal, eager to tell them about her trip east. "So it was the rain shadow I was tracking from the Cascades across the high plateau to your ranch." She pointed to her drawings of trees sketched along the way. "See how the Douglas firs changed to ponderosas?"

Contour lines connect points of equal elevation. If you follow one contour line, your elevation stays constant. If you move across contour lines, you're going either uphill or downhill.

Ricky remembered the differences, too. "Don't you love the smell of ponderosas on this side of the mountains?"

"For sure. I remember you'd told me about getting a feel for the east side by its fragrance. First the pines, then I caught a sour juniper smell at lower elevations and—"

"And scent of sagebrush where things get really dry," Ricky chimed in.

"And finally, smoke!" Ellie concluded.

Tony laughed. "You two are quite a pair. I think the fire tower near here might give you a great aerial view of what you saw. Want to check it out?"

"Absolutely," Ricky replied, amazed at how much his older cousin could do. He could drive a truck and ride a horse; he

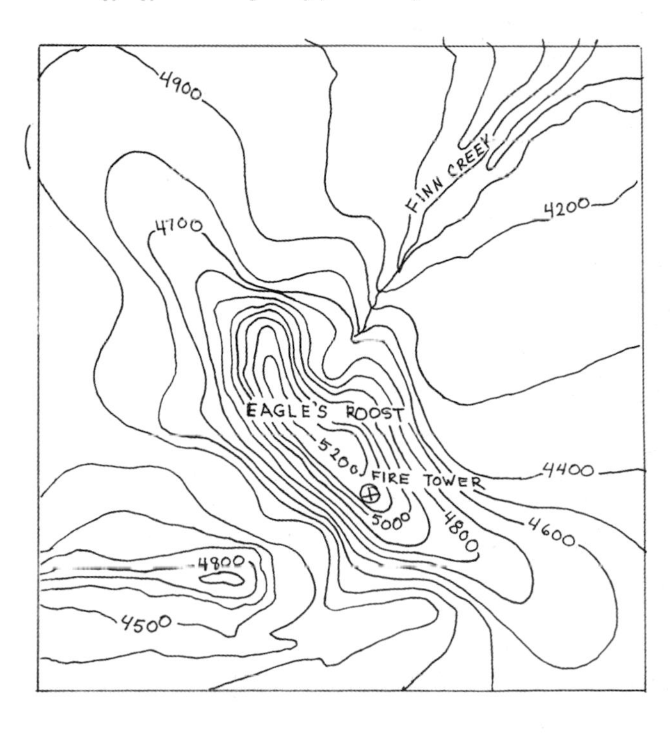

knew things about rain shadows and fire towers. And he always had great ideas.

After lunch Tony drove the gang up the Forest Service road to the tower. "The tower's just beyond those trees," he told them as he parked the van at the pullout. "Don't forget the binoculars, Rick."

Ponderosa pine trees are easily recognized by their tall, thick trunks and rusty-orange bark. When warmed by summer sun, the bark smells like butterscotch.

"In my pack," Ricky answered as they piled out of the vehicle.

Tony led them down a path, through a mixed forest of ponderosa pine and scattered Douglas firs. Ahead of them was a patch of light at the clearing around the tower. "There it is—the Eagle's Roost Lookout Tower," Tony announced. Ricky bounded for the tower stairs. Ellie and Sarah hurried to catch up.

When they reached the top of the tower, Susan, the forest ranger on duty, greeted them. "Welcome to my fire tower! Take a walk around," she said with a broad wave of her hand. In the center of the tower was an extra-large room, wrapped on all sides by glass windows. They saw Susan's desk, a map table, a ukulele resting against a big chair, her spotting scope, and other equipment. It was a snug living area. They walked slowly around the wide porch that circled the tower to catch the spectacular view. All around them was the fragrance of pines and firs.

"Wow—feels like you have a 360-degree view of the whole east side," Ricky said.

"Well, maybe not that far, but I can see more than a hundred miles across many of the valleys and mountaintops around here. Come inside, I've got a great model of this landscape to show you."

Ellie studied the three-dimensional map set in the middle of the room, then turned to look out the big windows. In the distance, she could see the peaks of the Cascade Mountains where she'd been with her dad. "If that's west, the sun will be setting on that side of the building," she figured.

"And the dense forest we see is north; the prairie is east," Tony added.

"Can you see the storms coming in?" Sarah wanted to know.

"Sure can—one of the most interesting parts of my job," Susan said. "Usually they come from the west, rolling in with big thunderclouds."

"Why are there so many thunderstorms on this side of the mountains?" Ricky asked.

"Good question. They start with dry, cool air flowing over the Cascades. On the east side, the hot air rises from our plateau. It's filled with moisture that comes from plant transpiration. That hot wet air is drawn up into the cool air from the west," Susan swooped her arms in a great arc above her head, "and creates an updraft. In big storms the hot air rises way high, more than forty thousand feet up; then a great cumulonimbus cloud forms."

Ricky remembered the big clouds and stormy weather that followed him on his trip to the ranch. "And then it gets windy?"

Susan nodded. "Sounds like you've been there. When hot air hits cold air, icy droplets fall from the top of the cloud in a rush. That cold, moist downdraft drives strong winds in front of the clouds.

Small winged seeds released from ponderosa pine cones can fly on the wind for hundreds of feet before settling to the ground.

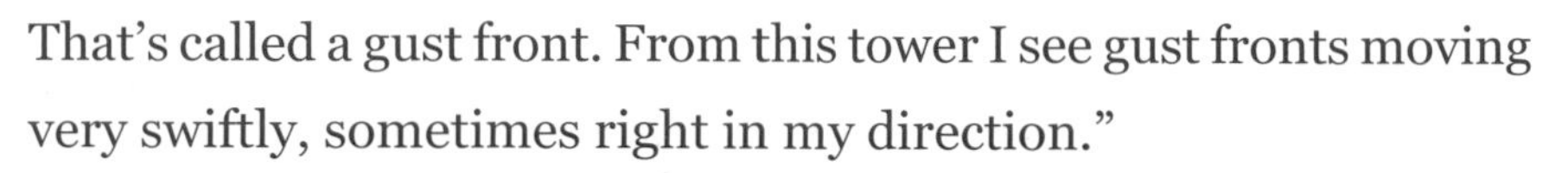

That's called a gust front. From this tower I see gust fronts moving very swiftly, sometimes right in my direction."

"I must've been riding a gust front on our way out," Ricky decided. In the distance it looked like clouds were forming. "Are those clouds coming this way a gust front?"

"No, I don't think so. They're not tall enough—and they're not moving fast enough. Y'know, there's a lot of help from satellite imagery and communications these days. Many of the old fire towers are mechanized. But I think my observations, right here on the spot, are important, too."

Tony considered the view closer to them. "Over there, see the Coyote Creek Fire site? There's still some smoke."

That caught Ellie's attention. "Look at all that blackened forest," she said. "Was it scary seeing it blazing from up here?"

"My job is to watch for surprises. Thank goodness, this one has been uncomplicated," Susan told her. "At first we were worried about a few ranch houses in that area, but it's almost completely contained now. Soon the crews will be working on the smoldering bits so it doesn't whip up again with the next big wind."

Sarah pulled out her camera and took photos from each direction on the tower with several extras in the direction of the fire. Ellie was watching two dark brown hawks spiraling in circles below the tower. "Can I use the binocs, Ricky?" With the binoculars she saw the bright red-orange of their tails. "Those are red-tailed hawks."

"They're riding a thermal draft up toward us," Susan explained.

Ellie was really enjoying their viewpoint. "We're actually looking down on them," she said. "They look so powerful, but graceful—slow and steady."

Ricky turned his attention to the landscape model again. He moved his hand across the ridges of the mountains, dips of the canyons, and flat spaces of the valleys. "Hey, Tony, here's where we stopped for ice cream, and there's your alfalfa fields, and Henry Hixon's ranch," he said, pointing to the landmarks he could identify. "But look, there's other kinds of grasslands, some scrubby juniper forests. Way out there—what are those greenish hills?"

"I know!" exclaimed Sarah, eager to have information to share. "That's where the fossils are."

Red-tailed hawks are among the most common hawks in North America. Primary feather "fingers" at the tips of their wide wings help them catch a "lift" from thermal updrafts.

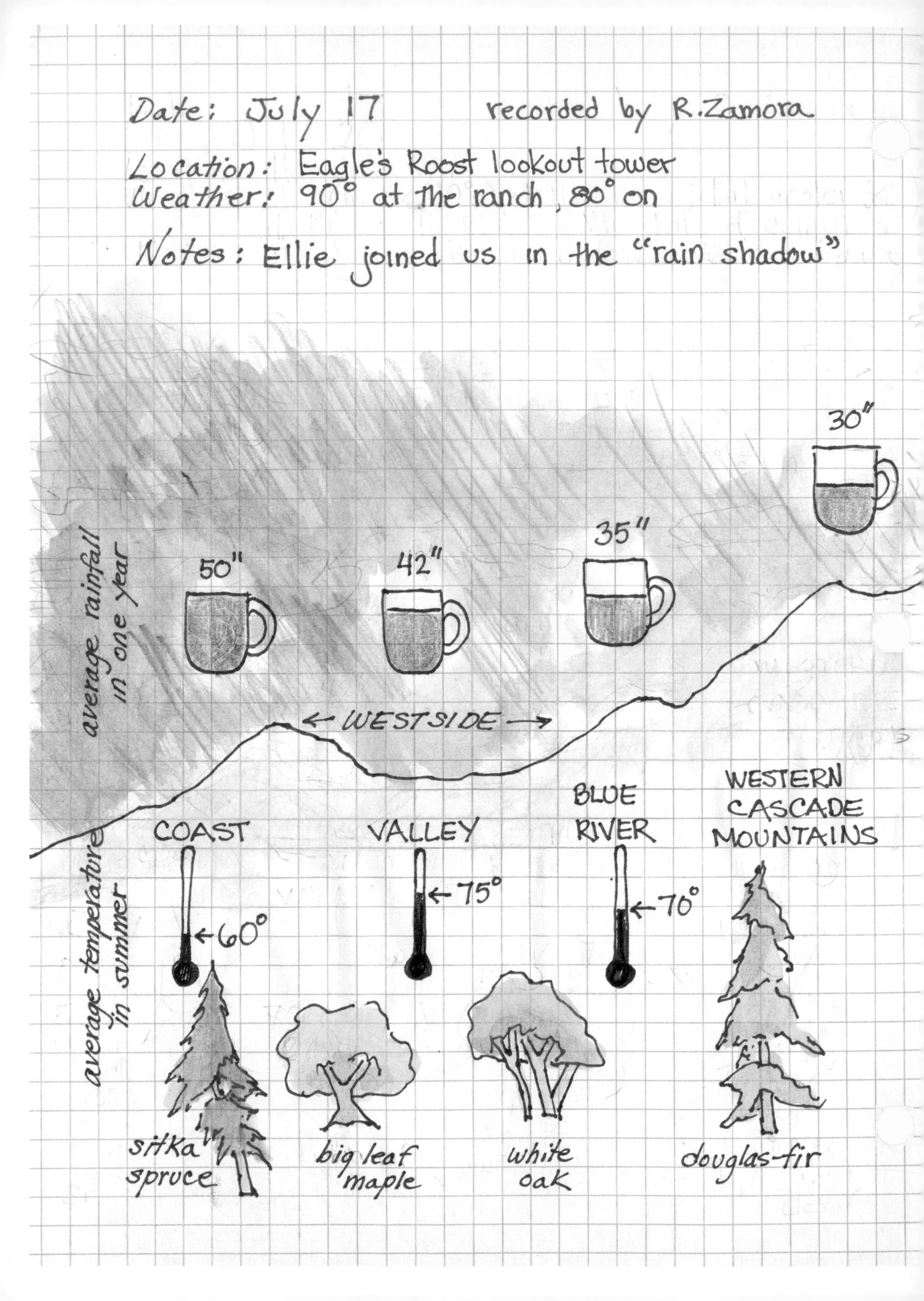
Date: July 17
recorded by R.Zamora
Location: Eagle's Roost lookout tower
Weather: 90° at the ranch, 80° on
Notes: Ellie joined us in the "rain shadow"
30"
35"
50"
42"
average rainfall in one year
← WESTSIDE →
WESTERN CASCADE MOUNTAINS
BLUE RIVER
COAST
VALLEY
average temperature in summer
←75°
←70°
←60°
sitka spruce
big leaf maple
white oak
douglas-fir

Ellie's Idea of a RAIN SHADOW
15"
10"
← EASTSIDE →
HIGH CASCADE MOUNTAINS
LAVA PLAINS
HIGH DESERT
65°
80°
90°
white fir
ponderosa pine
western juniper
sagebrush
bunchgrass

CHAPTER 4
Ancient Times

"Fossils?" Ricky and Ellie said in unison, eyes wide.

Tony laughed at their enthusiasm. "Well, if you like, we can take that route home," he said. "Time to go." They stepped out onto the porch, gave Susan an appreciative wave, and started down the long stairs to the ground.

Tony drove downhill until the road met the river, then followed the highway toward the fossil beds. On their way, they stopped at a road pullout to look at a landscape of spectacular red- and orange-striped hills. "These are the Painted Hills," Tony explained. "Millions of years ago volcanoes spewed ash that formed those layers of colored earth."

"And over there is Sheep Rock," Sarah added, pointing to a turban-shaped peak to the east. "It's volcanic, too."

"How about a peek at the peak?" Ricky asked with a sly grin on his face.

"Couldn't resist that pun, eh Rick?" Tony replied. "Sure, let's do it."

Near Sheep Rock, the trail led them into an extraordinary landscape. "Whoa, something's really different about this place," Ricky whispered. Erosion over millennia had cut into the rocks, uncovering whitish and greenish clays. It was a barren world, a little like a moonscape. Their boots made deep impressions in the soft soils that might have been formed thirty million years ago!

Black-tailed jackrabbits are hares; they have longer ears and taller hind legs than rabbits, and they can jump 10 feet in the air.

They'd been walking awhile when they came upon a marker. "This is where they found one of the fossils at the museum," Tony explained.

Ellie bent down to touch the rocks. “Imagine finding something that lived millions of years ago in these rocks,” she said. “Do we have time to visit the museum?”

“Sure,” Tony responded. “But we'll need to be kind of speedy because it closes soon,” he said, watching them turn abruptly and sprint toward the trailhead.

“One hundred species of mammals from these hills,” announced the sign at the entry. “The oldest fossils here are forty-five million years old!”

“Check this out!” Ellie exclaimed, looking across cases and cases of fossils, grand murals of ancient times, bones and models of prehistoric creatures.

Ricky was admiring a fossil with really big teeth. “A *Pogonodon*—kind of like a sabre-toothed tiger,” he noted from its label. There was an early rhinoceros, horses, and all kinds of ancient mammals.

Then Ellie came across a surprise. “Huh? Here are fossils from watery places—a turtle, a crocodile. How could that be out here in central Oregon?”

Rimrock caps on hilltops are the hardened remains of lava that flowed across the landscape several times during the last 50 million years.

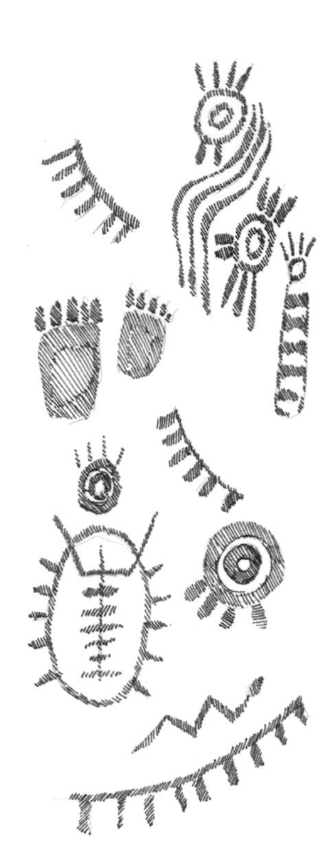

"There used to be an ancient lake and lots more water here," Tony told her. "See this aquatic plant fossil? It's from a place in the Painted Hills."

Ellie studied the Painted Hills stratigraphy, which looked like layers of cake made of orange and red rocks. "Were the oldest layers on the bottom?"

Tony nodded. "Yep, as long as the geology hasn't been disturbed from earthquakes or continental uplifts, oldest is deepest." He paused a moment. "Now, there's one other stop where we can see some very old records. I'm hoping they'll show up in this late afternoon light."

"Just a little longer here," Ellie pleaded. They completed their quick tour of the fossils and loaded back into the van. After driving a short way Tony parked near a narrow river canyon. "Follow me," he told them. "Stay tight against the rocks on this trail."

Pictographs are ancient drawings painted hundreds of years ago by Native Americans. The red pigment is made from crushed iron oxide mixed with something sticky, like blood, fat, or plant juice.

On the narrow trail many feet above the river, they scrambled around big boulders that had split off from the rock face many years ago. Ellie followed close behind Tony, wrapping her arms around a particularly large boulder as she made her way around it, scooting her feet on the narrow foothold. On the other side, Tony stretched out his hand to help her balance. When everyone was gathered, Tony pointed at what they'd come to see. "Pictographs—drawn hundreds of years ago by hunters coming through!"

Tilting their heads at a certain angle, they discovered dark reddish brown images drawn on the rock face. Some were just handprints;

others looked like simple symbols of unknown meanings. And there were stick drawings of animals; a couple seemed to be deer with antlers. "These look like records of hunting successes, or maybe just a record of people travelling by," Ellie said. "What a place to make their mark."

Shelled creatures called "ammonites" lived in warm seas millions of years ago. Their fossils are found in sedimentary rocks that began as ocean deposits.

"Mm, this must've been a distinctive landmark," Ricky noted after surveying their unique location. From their vantage point in the canyon they could see down the river heading toward Sheep Rock, and upstream the river made a definite bend coming out of a broad river valley. "An easy spot to recognize from far away—a good place for leaving messages," he concluded.

Ellie gently brushed the wall of rock with the pictographs. "Ancient people, leaving their sign right here. And nearby, really old fossils of animals in ages long before there were any humans here. Totally awesome."

A marsh rhino that died 40 million years ago is part of a mural at the John Day Fossil Beds National Monument.

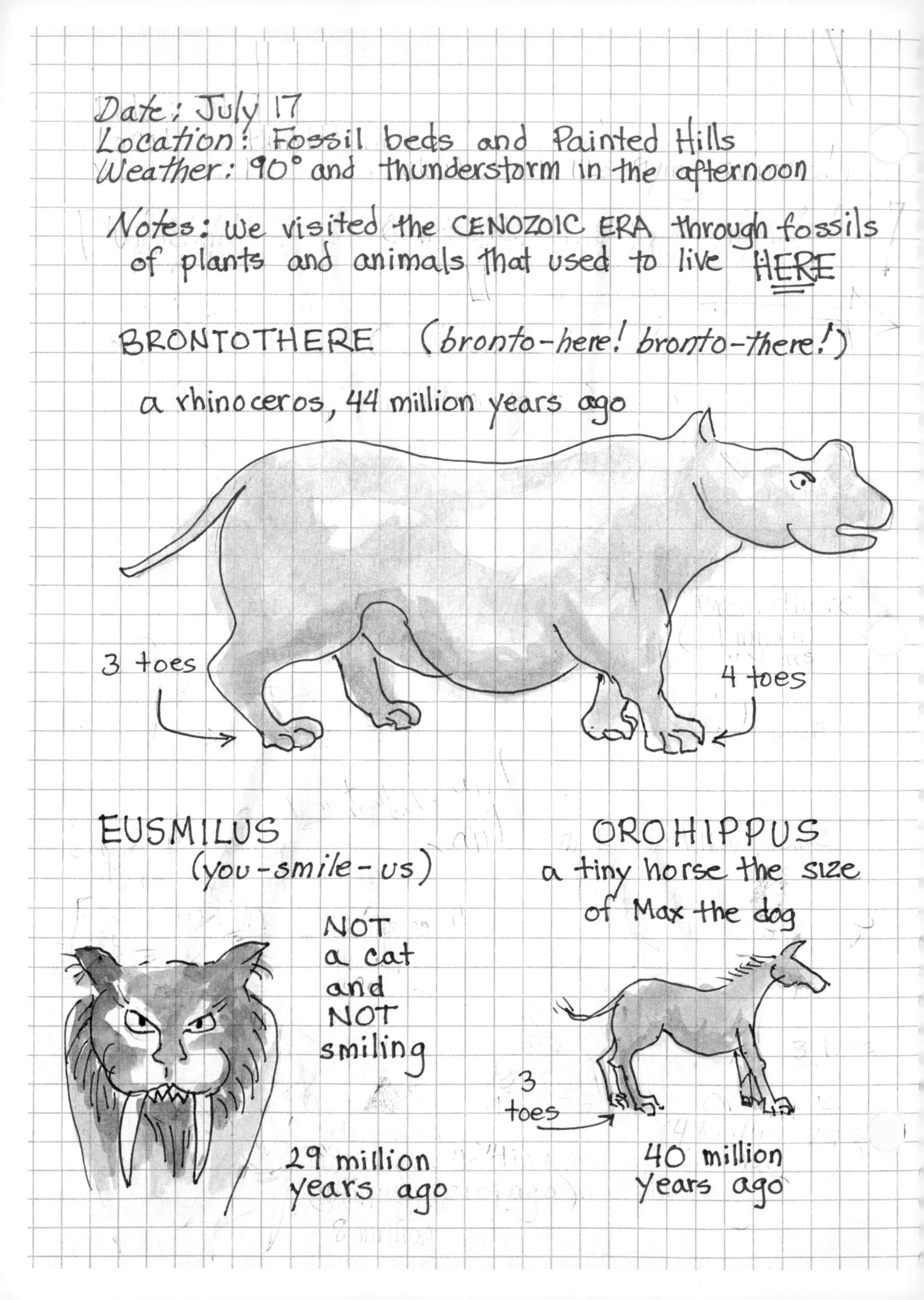

Date: July 17
Location: Fossil beds and Painted Hills
Weather: 90° and thunderstorm in the afternoon
Notes: we visited the CENOZOIC ERA through fossils of plants and animals that used to live HERE
BRONTOTHERE (bronto-here! bronto-there!)
a rhinoceros, 44 million years ago
3 toes
4 toes
EUSMILUS
(you-smile-us)
NOT a cat and NOT smiling
29 million years ago
OROHIPPUS
a tiny horse the size of Max the dog
3 toes
40 million years ago

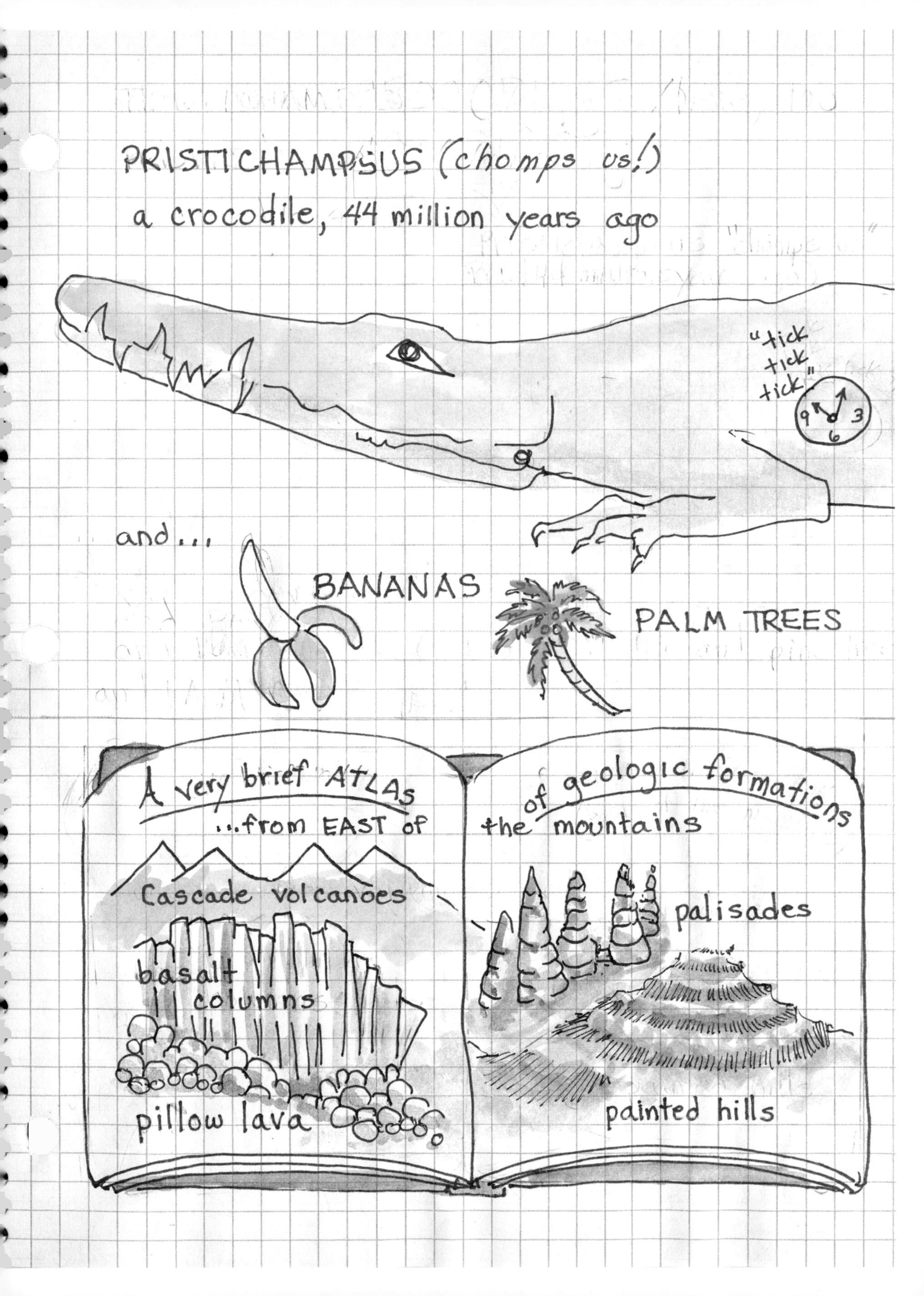

PRISTICHAMPSUS (chomps us!)
a crocodile, 44 million years ago
"tick
tick
tick"
9
3
6
and...
BANANAS
PALM TREES
A very brief ATLAS
...from EAST of
of geologic formations
the mountains
Cascade volcanoes
basalt
columns
pillow lava
palisades
painted hills

CHAPTER 5

Not So Long Ago

The next day Tony had promised a visit to Henry Hixon's ranch, but he was busy with chores all morning. So Ricky, Ellie, and Sarah were taking a break with Aunt Rosa and Ricky's mom at the house. As Ellie placed her glass of orange juice on the coffee table, she spotted a small clay bowl filled with what looked like unusual rocks. "What's in this bowl?"

The Paiute People of the Great Basin make tightly woven baskets for many purposes, like gathering food or carrying water, precious resources in these arid lands.

"We've collected those from our ranch and up in the mountains. You can pick up a piece," Aunt Rosa told her. "But be careful."

Ellie chose what looked like a piece of broken black glass from the bowl. "They're sharp!"

"The shiny black pieces are obsidian," Aunt Rosa explained. "Those chalky white ones are chert, and they're all volcanic rocks. Native Americans shaped them into arrowheads and spear points for hunting game, like elk and deer." Ellie continued to finger the obsidian's smooth flat surface, avoiding its sharp edges.

"If these were hunting grounds, was there a village near here?" Ricky asked.

"No, for centuries before settlers arrived, Native peoples traveled in seasonal rounds every year. There were traditional routes—kind of like hunters have favorite spots these days," Aunt

Rosa replied. "The men often hunted in open meadows where deer and elk grazed on shrubs. Around here, meadows probably stayed open because of natural fires, but in other regions Native Americans maintained grasslands by burning."

Ricky's mom had heard about the traditions, too. "Women gathered roots and berries, like balsamroot and huckleberries, in the same grasslands," she said. "Just imagine—they travelled as small family groups to hunt, harvest roots, and fish."

"Fish for salmon like we see spawning in the fall?" Sarah asked.

"Right, they followed salmon, trout, and even lamprey migrating upstream from the Columbia River. Fish migrated, berries ripened, animals moved with the seasons, and so did they. Before winter arrived they'd travelled back toward villages on the Columbia."

Sandals, made from twisted sagebrush bark 9,200 years ago, were found near Fort Rock, Oregon, buried in ash from the explosion of Mount Mazama.

"Making those rounds every year," Ellie mused. "That's a lot of trekking. Bet they worked hard at remembering their routes."

"Probably had maps in their heads about special landmarks, shallow places to cross rivers, good places to hunt, special spots for berries or roots—and pictographs like the ones we saw on the river canyon walls," Ricky suggested.

Aunt Rosa nodded. "Those were records from even earlier peoples—at least many hundreds of years ago. Keeping the memories of places with traditional stories was important to Native Americans."

"I think it's still important for us," Ricky's mom agreed. "Think of our annual visits to your grandparents in Mexico every year."

Stomp, stomp. The sound of boots on the porch. Tony was back after completing his chores. "Hola," he said. "How about going over to Henry's?"

"In a minute," Ricky replied as he, Ellie, and Sarah ran to grab caps. Before following Tony out the door, Ricky turned to his aunt. "Muchas gracias for the stories."

"De nada, don't mention it," she replied as she watched Max the dog leading them to the Hixon ranch.

Max barked to announce their arrival at Henry's house. "Hey, Henry, are you home?" Tony called as he bounded up the front steps. They heard Henry push back a chair and walk slowly across the wooden floors to greet them at the door.

The old cowboy peered through the screen door and greeted them with a wide grin. "Howdy Tony, Sarah. Brought me visitors?"

"Yep—this is my cousin Ricky and his friend Ellie. Remember, I told you Ricky arrived during the storm?"

"Of course. Come in, come in. You too, Max."

They stepped through the entryway into a large living room where sunlight filtered in through big windows. Ricky and Ellie looked with wonder at the gallery of photographs and pictures hanging on Henry's walls. Right away Ricky was attracted to an old map in the middle of the display. "Old River Valley, 1860," he read. Carefully drawn lines traced the course of the river and outlined the mountains in the southern and eastern parts of the valley. "How much has changed since this map was drawn?" he asked himself.

Ellie found a picture with an unfamiliar name on a storefront. "Shaniko"—where's that?"

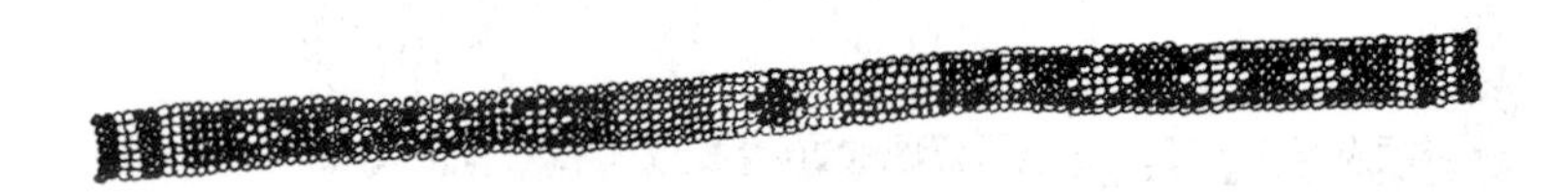

"In the late 1800s, lots of folks raised sheep. Lots of sheep. That's my grandpa with some of his flock at the Shaniko train station."

She studied the old photo of a hillside covered with sheep behind Henry's grandfather. His wooden wagon was stacked high with sheep wool and his horses looked ready to pull the wagon.

"Nowadays it's mostly cattle grazing around here, and Shaniko's a ghost town," Henry explained.

Ricky was looking closely at some photographs next to the map. "These photos look like they're the same hillside. Trees show in the bottom picture, but no trees in the top one. Were trees cut down before this top photo was taken?"

Henry eased into one of his big chairs. "They're the same hillside, alright," the old cowboy chuckled. "But that top photo was in my

Before-and-after photographs can show changes over time, such as the invasion of juniper trees into a sagebrush grassland.

Cutting invasive juniper trees can restore habitat for sage grouse, pronghorn, and other native animals.

grandfather's collection—he took it sometime before 1900, when there was just grass on that hillside. Then I stood in the exact same place and took the other picture a hundred years later! By then, the hillside was dotted with western juniper trees. Now, look out that window. What do you see?"

Western juniper trees get twisted and gnarly as they get old; they can live for more than 200 years.

"It's the same hillside again!" exclaimed Ricky.

"Yep," Henry smiled. "I like photographing how things change at photo points."

"Photo points?" Ellie asked.

"Yep, specific places where photographs are taken over and over again, for a long time," Henry explained. "Like time travelling."

"I could be a time traveller, too," Sarah said. "I'll take pictures of Coyote Creek every year and watch how things grow back after the fire." Henry gave her a thumbs-up.

Ellie was still studying Henry's photos. "Henry, how'd your hillside change so much over those hundred years?"

He sat back in his chair and looked out the window at the grass-covered hillside. "There used to be lots more wildfires here. Maybe every ten or fifteen years a good one would burn through, clean up the dry grass and dead wood. Those fires kept the western juniper out. That's when the first picture was taken."

"In some places there are ancient, three-hundred-year-old junipers. But where there's more moisture, there may be cycles of years when junipers do better than in others; then they don't make it to old age. One thing for sure, juniper trees don't survive fire well. When we didn't let the natural fires burn in our valley, junipers started popping up all over. If fires got started, from lightning or whatever, we'd put 'em out. That's how it was when I took the second picture, with junipers dotting the hillside."

"But there aren't any junipers there now," said Ricky, pointing at the grassy hillside beyond Henry's window.

Henry nodded. "Junipers have long roots and they hog a lot of water. Lately, we've been cutting them to conserve water. The grasses are doing pretty well. It looks a little like how my grandfather saw it, don't you think?" Ricky agreed.

"I saw all kinds of forests driving out this way," Ellie said. "Are those trees all sensitive to fire like juniper?"

"No, not at all," Henry said. "Higher up in the mountains, I think you passed through some tall and skinny lodgepole pine. That's a tree that really loves wildfire. Fire opens its cones and releases its seeds. It won't even sprout without fire."

"The fire-resistant trees you see around here are the ponderosa pines," he went on. "They grow on lava lands east of the Cascades. They've got tough, thick bark that protects them from burning. Those big old trees are almost fireproof. Until about a hundred years ago fires came through these forests every few years and cleared out the underbrush. The early settlers used to say you could drive a buggy between those big ponderosas."

"Whoa," said Ellie. "You couldn't drive a buggy through the Douglas fir forests on the west side. You'd go about three feet and get stuck in all the brush!"

"So you can imagine what happened without fire in Ponderosa forests every few years—trees grew closer and closer together."

"Not a good thing?" Ricky asked.

Henry shook his head. "One thing led to another. Trees competed for water and space. Sometimes bugs infected unhealthy trees. Dead trees fell over and made wood ladders for fire to climb into the tops of the forest. So, the fires got bigger and more destructive."

"What happens when there's a fire now?" Ricky asked.

"Well, take the Coyote Creek Fire, for example. Firefighters monitored the fire and kept it away from homes and ranches. They let it burn in places to clean out the forest and keep the dead wood from piling up. It's not quite like the horse-and-buggy days, but we're learning to live with fire in these forests. We're lucky the Coyote Creek Fire didn't last too long."

"For sure," Tony answered. "Reports sound good right now."

"Before you go—why don't you take this map of our valley? It shows the hills we've been talking about and a natural prairie where I set up photo points. Maybe you can help me catch a few new pictures."

Ricky quickly surveyed the map then carefully rolled it up. "Sounds like the kind of job we like!"

"And I've got my camera for photo points," Sarah added.

"I spent time in those hills learning about the prairie restoration projects last year," Tony said. "And now I'm in charge of a couple of monitoring sites. So we'll find a way to get there." He stood up to go. "We'll give you a report in a few days, Henry."

The kids took a short cut across the upper pasture on their return to the Zamora's ranch. Cows in the lower pasture raised their

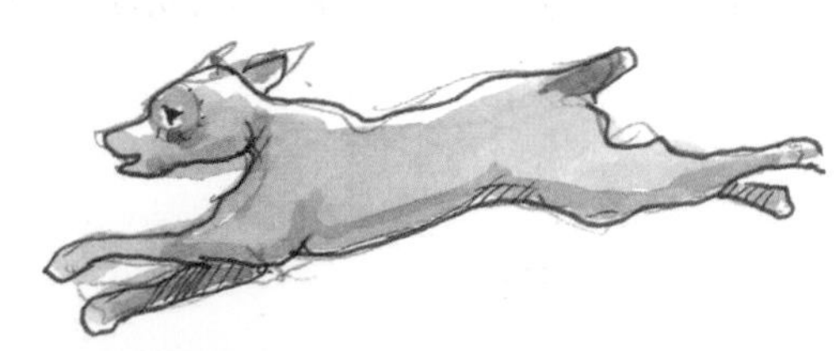

heads, took note of the newcomers in their field, and began sauntering toward them. "They're just curious—no chance they'll catch up with us," Tony reassured them. Just the same, everyone stepped up the pace so they reached the barbed-wire fence separating the two ranches ahead of the cows. Tony put his foot on the lower wire so Ricky could squeeze between the strands. Ellie watched as Ricky bent low and got to the other side without a scratch. She crouched down and followed him. On the other side Ricky held the barbed wire apart for Sarah, then Tony. Max slipped through, no problem, and trotted ahead.

The western rattlesnake's scientific name, *Crotalus*, comes from a Greek word meaning "little bell," which refers to its rattle. If discovered in a hiding spot under a rock or in the tall grass, *Crotalus* often use their rattles as a warning.

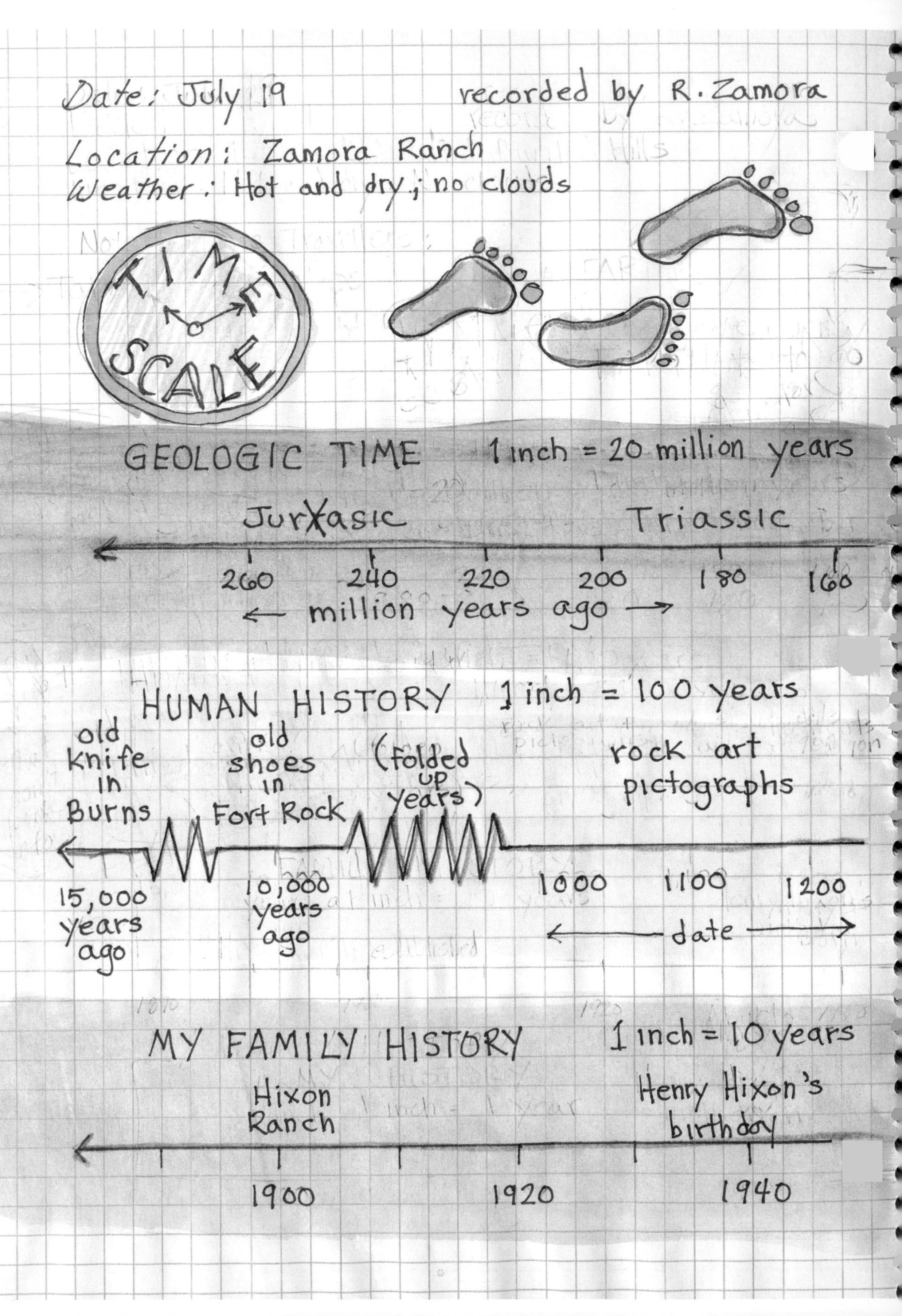
Date: July 19
recorded by R. Zamora
Location: Zamora Ranch
Weather: Hot and dry; no clouds
TIME SCALE
GEOLOGIC TIME
1 inch = 20 million years
Jurassic
Triassic
260
240
220
200
180
160
← million years ago →
HUMAN HISTORY
1 inch = 100 years
old knife in Burns
old shoes in Fort Rock
(folded up years)
rock art pictographs
15,000 years ago
10,000 years ago
1000
1100
1200
← date →
MY FAMILY HISTORY
1 inch = 10 years
Hixon Ranch
Henry Hixon's birthday
1900
1920
1940

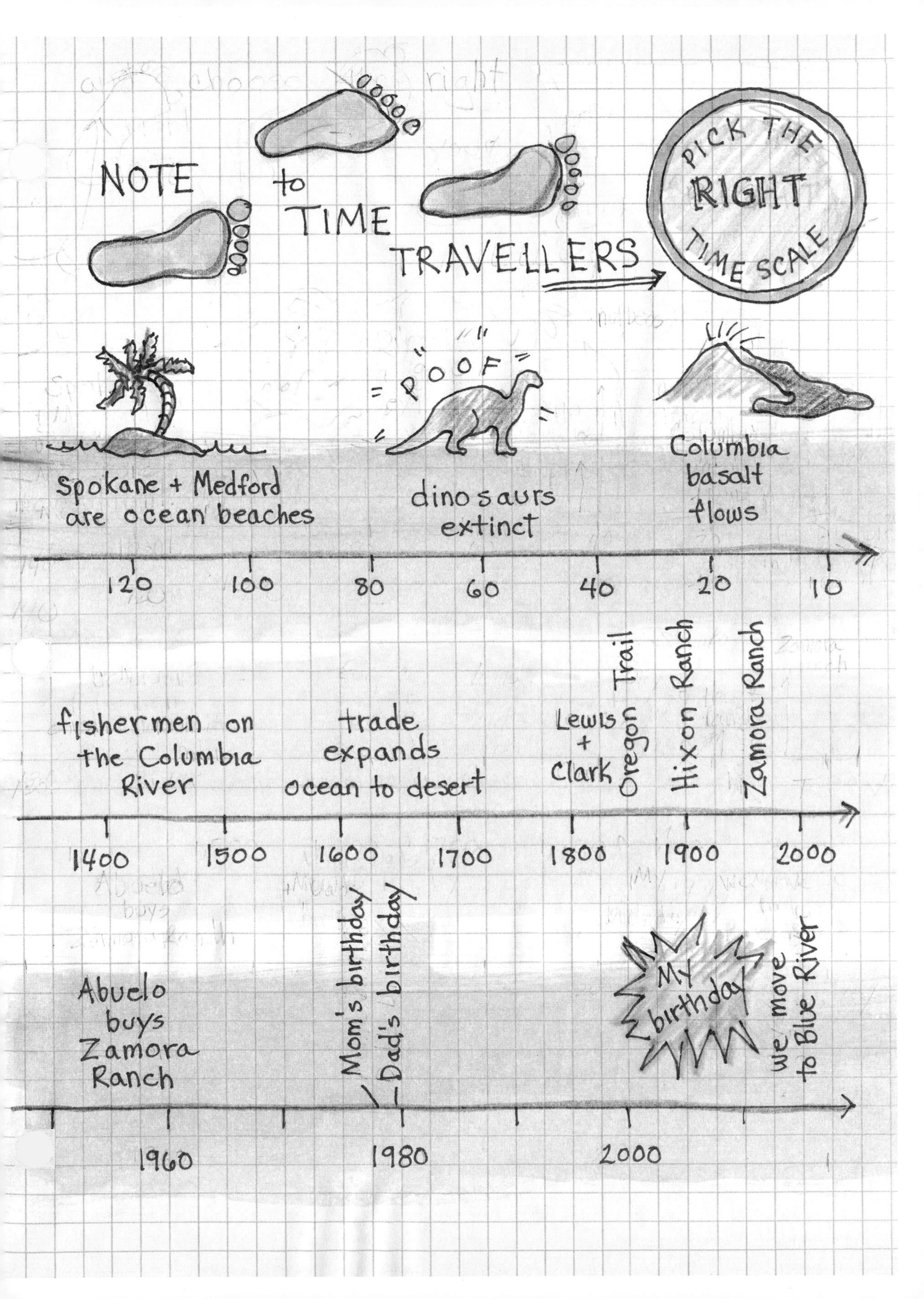

NOTE to TIME TRAVELLERS
PICK THE RIGHT TIME SCALE
POOF
Spokane + Medford are ocean beaches
dinosaurs extinct
Columbia basalt flows
120
100
80
60
40
20
10
fishermen on the Columbia River
trade expands ocean to desert
Lewis + Clark
Oregon Trail
Hixon Ranch
Zamora Ranch
1400
1500
1600
1700
1800
1900
2000
Abuelo buys Zamora Ranch
Mom's birthday
Dad's birthday
My birthday
we move to Blue River
1960
1980
2000

CHAPTER 6
After Fire

"Hey, El," Ricky called. "Is that your dad's truck coming up the drive?"

"No, can't be. He said he was gonna be busy all week."

But there was Ellie's dad, Derek Homesly, hopping out of the truck, coming up the front steps.

He gave Ellie a big hug. "Hi guys," he said. "I'm off to check out the Dead End Trail fire site today. A big fire from last year that's just beginning to recover. How about coming along?"

"I was hoping we could see the Coyote Creek fire," Ricky said a little hesitantly.

Ellie's dad nodded his head sympathetically. "Understood," he said. "Problem is, even when a fire is technically contained, it's still a very dangerous place. Right now they're looking for hot spots. And with all those charred trees, who knows which trees will topple? Forest Service crews will be working at making it safer for weeks. Meanwhile, the Dead End site needs survey work. I thought you could help me."

"I remember that fire," Tony said. "It was bigger than the Coyote Creek fire, but farther away from us."

Ellie was already putting on her hiking boots. "I'm in," she said.

Of course Ricky and Sarah were quick to agree. "My truck won't carry us all," Ellie's dad told Tony. "Could you drive us in your van?"

Tony grinned. "I'm your man." Ellie helped her dad transfer clipboards, measuring tapes, and monitoring gauges into the rig. "Let's take the highway south and catch the Forest Service road over the ridge," her father directed.

Hairy woodpeckers look like little soldiers at attention. You can attact them to your garden with suet feeders.

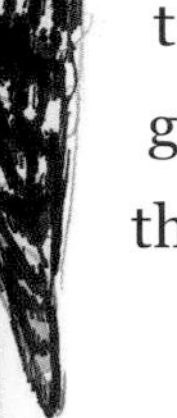

On the bumpy gravel road leading to the fire site, sunlight filtered through the ponderosa forest. Little patches of light on the dry forest floor were littered with long pine needles. "No signs of fire here," Ellie observed.

"But it looks pretty dry," Ricky noted. "Lightning might start a fire in a flash."

Ellie's dad agreed. "For sure. Summer dryness, summer heat, and thunderstorms add up naturally to frequent fires." When they reached a broad meadow, he said, "This is the spot, according to the map."

Tony noticed deep tire tracks in bare spots as he parked the van. "Signs of fire vehicles from last year, I'd bet," he said.

Not a sound in the air as they piled out of the van. No trees to catch the breeze, no insects chirping. Tony charged up a small knoll leading away from the meadow, and Ricky followed, hurrying to keep up with his cousin's long strides. "Look at that!" Tony exclaimed.

Burned forest stretched as far as they could see. There were skeletons of shrubs, small ashy piles of burned plants, and young shoots of leafy green ones. Spikes of grasses poked out of the blackened soils. Everywhere downed, burned trees lay as charcoal logs on the ground. Other trees, still erect, had been burned to black trunks bare of any branches.

But there were stately ponderosa pines still standing. Many had black fire scars extending up the tree trunks, and crispy brown needles. High in the upper canopy, though, the ponderosas were still green. "Looks like those ponderosas are something special," Ricky said.

"Proof that ponderosas are truly adapted for surviving fire," Ellie's dad replied. "The thick bark of the older trees protects them from fires on the ground. And it shields the inner, living parts of the tree trunk."

"Are they the only trees like that in these forests?" Ellie asked, remembering their conversation with Henry.

"No—another conifer in these woods, the western larch, also has a thick skin. But I don't see one right here." Tony and Ricky started across the burned forest floor, approaching one of the standing ponderosas. They couldn't resist sticking their fingers between the thick plank-like patches on the ponderosa bark. "Check this out, Ellie," Ricky called to his friend.

When Ellie studied the tough ponderosa bark, she was reminded of the forests where she lived in Blue River. "Pretty different from Douglas firs, I'd say."

"You've got that right, Ellie," her dad answered. "Old Doug firs can resist some fire, but bark of younger Doug firs aren't nearly as thick. They don't do well where there's frequent fire." He pointed to a cluster of trees that looked like huge charcoal stakes, burned to the very tops. "See over there—those were junipers. They really can't make it through big fire."

As they started hiking downhill, they heard a tap, tap, tapping. Looking up toward the tops of the pines, Ellie spotted a noisy, hairy woodpecker. "Look at him! I'm surprised there's anything living here."

"Aren't critters amazing?" her dad replied. "Some get away by fleeing from the site, others find a hiding place somehow."

"Wouldn't birds fly away?" Sarah wondered.

"True, many birds do leave, but woodpeckers might look for cavities high up in the trees. As soon as the fire cools, they're on the spot to find insects, especially beetles."

"Beetles like the ones we see in decomposing logs?" Ellie asked.

"Or like this speckled one with the long antennae?" Sarah asked, trying to grab it before it flew away.

"You're both right—bark beetles, pine beetles, long-horned beetles, the kinds of insects that can bore into wood," Ellie's dad responded. "Right after a fire there's lots of trees for larvae to chew on. It's a post-fire food web—first the insects, then the birds eating the insects, followed by raptors coming after the birds. And that's just what's going on in the treetops."

Fireweed is one of the first plants to appear after a forest fire or other disturbance. Its seeds can lie dormant for many years, waiting for the heat it needs to sprout.

"Quick, over here, Ellie," Ricky yelled. He'd seen a brush rabbit. It was easy to follow. "Looks like it's trying to find a place to hide."

"There's plenty of refuges—in the tree canopy, on the ground, and in the subterranean," Ellie's dad said as the rabbit tucked under a big charcoal log.

"Oh, subterranean," Ricky pondered. "Under a log would be a good spot for that rabbit, or maybe it has a burrow there? Anybody else go underground?"

Ellie's father grinned as Ricky stooped down to find the rabbit. "You'd be surprised at the number of animals that do that—snakes, lizards, badgers, even earthworms."

"Really didn't expect to see wildlife still living in the burned forest. Makes all this charcoal blackness feel less eerie," Ricky said quietly. He was beginning to form ideas about what kind of map to draw—maybe it'd be one about animal hiding places in the burned forest.

The kids followed Ellie's dad down the gentle slope until they reached a small stream. Ellie noticed a swarm of small flies. They were midges, emerging, bopping up and down like tiny dancers above the water. "Life goes on in the stream, too," she remarked.

Her dad reached into his backpack to start a special task. "This is where I need your help," he said. "See the colored plastic flagging I've put out on the other side of the stream? Those are survey lines that we call transects. We'll be watching changes along the transects for many years as this forest recovers from the fire. Right now it's pretty gritty with all the charcoal, but would you help me with a few transects?"

"Ready and willing," Tony said.

"Terrific. It's pretty basic. I'll give each pair of you a data sheet that has a transect number on it. Walk from the stream straight uphill toward the flag that matches the one on the stream bank. We set them out one hundred meters from the stream. Count the number of living trees you see, and how many trees with no signs of living needles, leaves, or stems."

"Want to know about the scars on the ponderosa pine trunks?" Tony asked.

"Mm, good question. They may be hard to measure—but take a guess at the height of the burn marks you see, and write them into the notes part of the sheet."

Tony and Ricky grabbed a data sheet and set out for the transect with black and yellow striped flags; it was farthest away. They

were soon collecting data. Sarah and Ellie weren't far behind; they took the transect with blue polka-dot flags. Derek Homesley walked one on his own. Because the burning had been patchy, with live and dead trees scattered around, the teams moved along the steep terrain at different speeds.

Tony and Ricky encountered several scarred ponderosas. Some scars were too high to reach. "Let's compare them to how tall we are," Tony suggested. On the first tree, they figured scars were well over six feet. Climbing was steep the next few meters. "Nasty," Ricky commented as he tried to dust off ash after scrambling over a burnt log. Tony was recording another tree scarred about five and half feet up.

Ellie and Sarah had to climb around several rocks. They tried not to trip in the gritty charcoal and ashen soils, but their hands got grimy. They'd counted only two living trees and were recording the sixth dead tree skeleton when a light thumping on the ground startled them. A young fawn and its mother bounded through the blackened forest and up the other side. "There's a pair that's returning early," Ellie commented as she entered their observation into the notes column.

Ricky and Tony found two trees with scars less than four feet high near the end of the transect, where the path became more level. They were brushing soot off their data sheets as Sarah and Ellie reached the top of their transect, too.

Ellie's dad, climbing over several downed trees, called to his helpers, "Come see what I've found sunning itself on this log." When they made it over

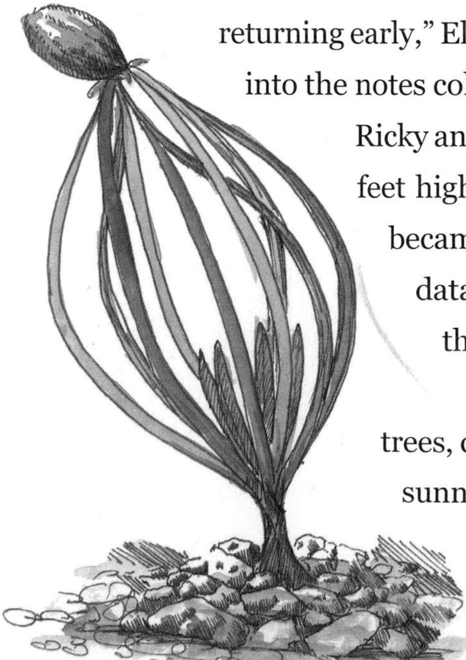

A lodgepole pine seedling gets a good start in soft soil.

to his transect, they saw a bull snake coiled in the bright sun. "This snake probably burrowed near here during the fire, and now it's out warming up."

"May not be there for long," Tony remarked, looking into the sky. "Scree, scree." A red-tailed hawk circled overhead. "The raptors are back, looking for prey. Seems food might be easier to see after the tree canopy is gone." Silently, the snake disappeared under the log.

Ellie's dad collected each of their data sheets and tucked them in a folder marked "Dead End Trail Fire, First Year Survey." Thanks everyone—you've made my job easy today. Now, let's head back and beat the heat."

The road skirted the edge of the fire as they left the site, then veered away from it going downhill. On the other side of the road, Ricky noticed a sudden change from messy shrub cover to forest that looked freshly cleared with trees still standing. "What happened here?"

"Good eyes, Ricky. Five years ago they did a small prescribed burn here."

"Does that mean fighting fire with fire?" he asked.

"That's right. We're lowering the density of plants and helping the ones that need fire to do well. There'll be more habitats for wildlife, and they'll get more kinds of food to eat."

"Looks kinda like a park, with lots of room to run around," Sarah observed.

Ellie thought about how dry the forest seemed. "Is it hard to know when to burn?" she asked.

"We look for the perfect time in the spring or fall, when plants are kind of wet, winds are calm, and temperatures cool. We have

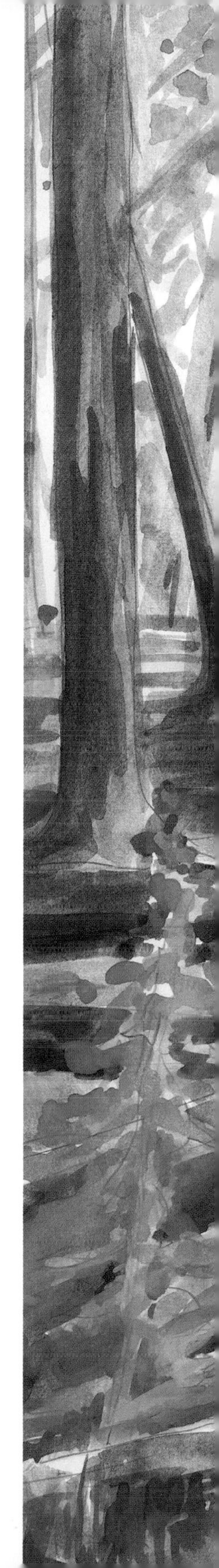

lots of helpers and control the burning by taking it very slowly. In this forest they wanted to burn the understory—the small woody bits, grasses, and brush that are easy to burn."

They got out of the van to look around. "What are those skinny, bright green conifers?" Tony asked, pointing to a distinctively different tree. "I've seen them in the high country when we're rounding up the cows."

"That's a larch, Tony. It's the other conifer with thick bark we were talking about earlier. One of the few conifers that sheds its needles every year. In the spring, new needles come back in brilliant chartreuse green. Young larch need bright light to grow tall, so the open patches made by burning help a lot."

"Look over here," Ricky said. "Huckleberry bushes—with berries that'll be ready to eat pretty soon."

"It only takes a year after fire before the berries get going," Derek told him, kneeling down to look at another young shrub. "And this is a wild lilac. It had snowy white blossoms in early summer. They're adapted to fire—they have seeds that can store deep in the soil. They can be dormant for more than a hundred years.

Lodgepole pinecones are closed up tightly, but open up after a fire to release their seeds.

"No way!" Ricky exclaimed.

"Yep—true story. They wait until hot temperatures from fire help them to sprout."

Near the wild lilac Ellie found a big, slightly prickly ponderosa cone. "Nice," she said, picking it up. "You can see the seeds inside."

Sarah found a smaller cone. "Here's a different kind—has a waxy coating, and its scales are really tightly closed."

She handed the cone to Ellie's dad. "I was hoping we'd find one of these," he said. "It's a lodgepole pinecone. Over there, see that tall pine with shorter needles than the ponderosa? They're called *lodgepole* 'cause the Native Americans used them for long, straight totem poles."

"Oh, right," Ellie said. "Henry told us about these cones. Fire pops them open."

"So plants have all kinds of fire-resistant tricks up their trunks, in their cones, and in their seeds." Ricky was putting the plant story together.

Sarah plucked a few scales from the Ponderosa pinecone. "For my notebook," she told Ellie.

On the ride home, Ricky remembered the smoky days and the fire's glow, the firefighting vehicles and helicopters overhead. But what he'd seen today was just as impressive. Maybe he'd draw a map of how the live and dead trees made a patchwork created by fire. And scurrying between the burnt trunks, coming out from under the big rocks, foraging on the living trees, would be the mammals, birds, and reptiles who survived.

Date: July 22
Location: Mr. Homesly's fire recovery site
Weather: Cirrus clouds in morning; hot afternoon

Notes: Transects help me read between the lines!

TOPO maps show UPs and DOWNs with CONTOUR LINES

(and point toward the NORTH ⇗)

RELIEF maps show UPs and DOWNs with SHADING

Lodgepole pinecones
=POP= open
like POPcorn!
mountain bluebird
ladder back woodpecker
Clark's nutcracker
mule deer
Some plants + animals do okay after a fire (as long as the fire is not TOO hot!)
huckleberry
fireweed
morel mushroom
FIRE
small fires can clean out thickets that CHOKE the forest
=gasp=

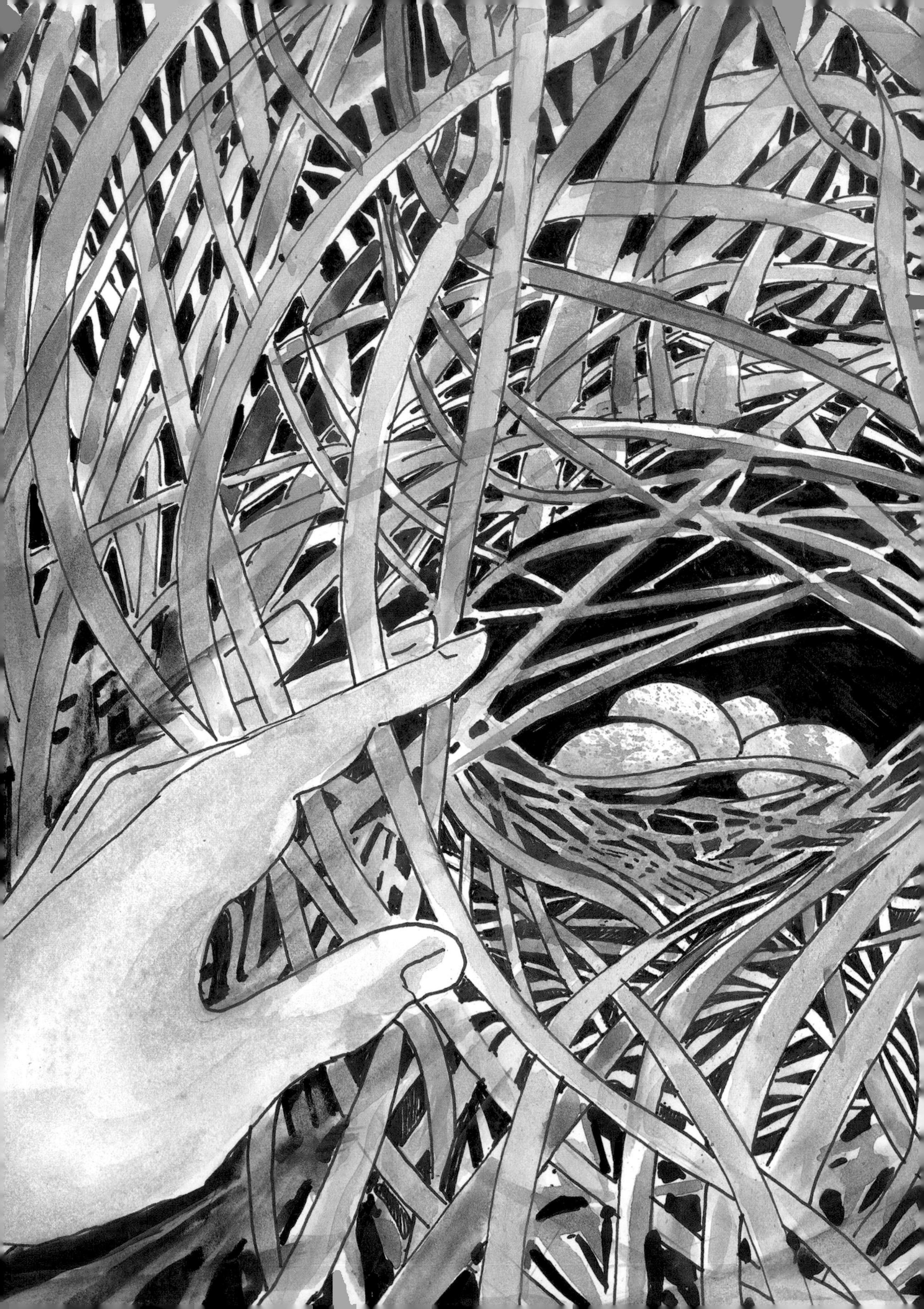

CHAPTER 7
The Prairie

"Way cool," Ricky declared. The day after their trip to the fire site, Ricky and Ellie finally got back to Henry's map. "It's a topographic map that shows us elevations. Check this out, Sarah. "

Sarah traced an imaginary route with her finger. "There's the river in blue, and this curvy line in red looks like the highway. But what are these squiggly grey lines?"

Meadowlarks are easier to hear than to see; you'll know them by their flutelike song and prominent yellow "V-neck sweater."

"We can figure out changes in elevation with those lines. Lines close together mean sharp elevation changes. See how these lines circle around Eagle's Roost Mountain?" His finger traced a small circle at the top of the mountain. "That contour line marks sixty-five hundred feet above sea level—and the line down the mountain is the five thousand feet marker."

Ellie studied the places where the contour lines curved around mountaintops, then looked where the lines were more spread apart in flatter lands. "Where we are, next to the river, we're even lower. Elevation's only thirty-five hundred above sea level here. We felt that difference travelling downhill from Eagle's Roost."

"Right. We're reading 'between the lines.'"

As Tony joined the crew he looked at the map over their shoulders. He pointed to a plateau on the northern border of Henry's ranch. "This is where I'm helping the wildlife biologists and ranchers restore the prairie. Sheep and cattle have been grazing around here for more than a hundred years. But on the high plateau the growing season is short and not very good for raising crops or livestock. It's more like the original bunchgrass prairie. "

"Bunchgrass. Let me guess," Ricky chimed in, "they grow in bunches?"

"Uh huh. Each plant makes a clump, a hummock—a bunch."

"Do Henry's cattle graze on bunchgrass?"

"They would if they could. Deer and elk like bunchgrass, too. Henry figures weeds and non-native grasses invaded more than a hundred years ago. To bring back that native prairie, they're doing prescribed burns and excluding grazing with fencing."

Ricky remembered their visit with Susan. "So there's prescribed burning not just in forests, but grasslands, too?"

"Right. Our monitoring records will track changes up there. C'mon, we can check it out today."

Tony drove them up the dirt road that followed Henry's fences, heading to a ridge on the northern border. When they reached the prairie, they stopped at a cattle guard gate. Sarah jumped out of the van, unhitched the heavy gate, and swung it open so Tony could drive through. After the van rumbled over the metal grate, Sarah heaved the big gate closed and got back in.

Ellie was surprised how smoothly Tony and Sarah executed their gate ritual. "Why do they have that kind of gate?" she asked.

"The metal grate keeps cows from getting in. The swinging gate is for restricting people. Both help protect these grasses," Tony explained.

When they stopped at a pullout, Ricky brought out Henry's maps. They looked for the valley where they'd been and the broad meadows they'd come to visit. "Here's our turnout," Ricky pointed out. "Where do you think we'll find Henry's photo points?"

After studying the map, Tony made a plan. "Let's walk toward that thicket of trees. Those aspens are part of the plot I'm monitoring. We'll catch one of Henry's photo points midway, then one up on the aspen ridge."

"Is this where they did the prescribed burning?" Sarah asked.

"Yep." Tony opened his arms, spreading them in a wide arc. "There've been burn patches in this meadow and uphill."

They walked across the golden field of grasses and found a narrow trail to follow. "Deer or elk have been through here," Ellie thought. She noticed that the lush grasses were growing back pretty fast after the fire. At a patch of blooming wildflowers, she and Sarah bent down to investigate.

"Seems there are bare spots around clumps of grasses. Good for wildflowers, maybe?" Sarah suggested.

"Probably so." Ellie was looking at perky yellow blooms that stood only three inches high. "These look like miniature sunflowers."

"Those are balsamroots, Ellie—blooming pretty late this year," Tony told her. "Glad to see that burning made some patchiness for flowers in the grass."

"Hey, Tony—over here," Ricky called. "I think there's a bird nest."

Ricky moved slowly through the tall grass, carefully raising each foot high in the air. Silently, he crouched down on his knees.

When he spread apart stems of grass, he discovered a dome-shaped tunnel about three inches high that formed an entrance to the nest. "I hadn't thought about birds on the ground," Ricky said very quietly. "Check out this amazing nest."

Tony nodded. "I've seen one of these at our ranch. It belongs to a meadowlark."

"Like the one I saw singing the other day," Sarah said. Ricky moved away so the girls could see, too.

"We should be quick—don't want to scare away the parents," Tony cautioned. Click. Sarah took a snapshot before she snuck away.

Farther into the meadow, Sarah found a big patch of bright yellow goldenrod, just starting to flower. "Ellie, come see these tiny bees," she said. "I don't think they're honeybees or bumblebees."

Ellie pulled a hand lens out of her pocket and gave them a closer look. "These are sweat bees. I remember their really small size; Mom told me there are many, many kinds of them. Can you see the little bits of pollen they're carrying on the upper parts of their legs?" Ellie gave the hand lens to Sarah. There were plenty of sweat bees foraging on the goldenrod, so it was easy to watch them. "Oh! A skipper butterfly, too." Ellie watched it flittering among the long stalks of goldenrod flowers. "Its yellowy wings almost blend in with the flowers."

When they stood up from their bee-watching, they saw Tony heading for an old cottonwood snag. "Here's Henry's photo point," he said.

Balsamroot is prized by Native Americans, who cook its thick, meaty root.

Sarah caught up with him in a hurry. "Let's try to get the same angle on the prairie as Henry's old pic," her brother suggested. "Kind of tricky; brush removal and burning have changed the prairie a lot. Whadda ya think?" he asked, handing her the older photo for comparison.

Sarah held the picture to find a good landmark. "Over there. See the sliver of road in the photo? I think that's the one we came down."

Tony smiled. "Excellent."

Click. One photo into the record.

The sun was bright and everyone was getting hot. As they stopped for a water break, a male meadowlark flew silently over the grasses then dove into the clump where they'd found the nest. Ricky was very pleased. "Perfect. Things back to normal."

They'd continued walking toward the aspen thicket when Ricky stopped again. "Check out the wingspan on those birds," he said, pointing high into the sky.

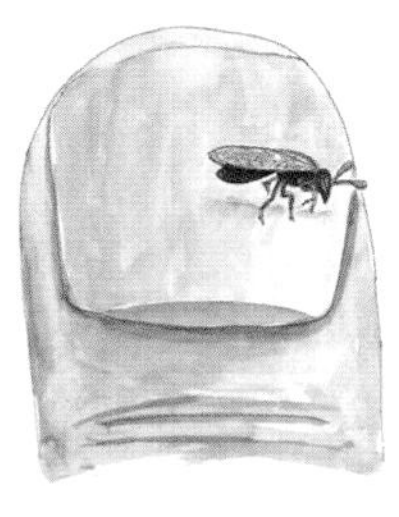

Sweat bees are often the most common bees in a habitat, but they're so small (0.1 to 0.4 inches) they're easy to overlook.

Ellie thought a moment. "Could be golden or bald eagles around here."

Ricky took a second look through the binoculars. "Looking for the color of their heads . . . any white? No, they're not adult bald eagles. If they were young bald eagles, they'd be pretty light-colored on their underbellies and wings . . . but they're really dark all over. I'd say they were goldens. Wanna see, Ellie?" Ricky handed her the binoculars.

She agreed with Ricky. "Nice. They're golden eagles for sure."

When they'd reached the end of the meadow, Tony announced, "Here's another part of the experimental burn—where they're trying to restore the aspen groves." He pointed to clusters of skinny young

The golden eagle is the largest bird of prey in North America, and it's also the national bird of Mexico.

trees on the hillside ahead of them. Their leaves shimmered like leafy coins in the sunlight. Ellie recognized the aspens by their shiny white bark. "Haven't seen many aspen out here," she said.

"That's partly because they only live at high elevations. But even here we're worried that there are fewer than in the past. We'll have to climb a bit to reach them."

Ellie surveyed the steep dirt trail. "No problem. We're tough." There was a little huffing and puffing as they zigged and zagged up the hill, but they reached the shady aspen grove in about fifteen minutes.

While everyone took another cool swig from water bottles, Tony explained, "Believe it or not, this grove is actually a clone from one root. It grows only by resprouting, but there hasn't been much sprouting lately. Deer and elk like to browse on the young aspen, and fire suppression has also been a problem."

"Do aspen need fire to resprout?" Ricky guessed.

Tony gave him a thumbs-up. "Excellent deduction. They burned this area three years ago to save the aspen."

In the small grove, everyone took a slightly different path winding around the slender trees. "Looks like the burn removed the underbrush. Not bad walking through here," Ricky noticed.

At a sunny spot Ellie stopped to watch several bumblebees. They were working over a tall plant tipped with spires of small white flowers. She watched them stick their fuzzy bodies into each blossom. "Bumbles are busy on this hummingbird mint," she said.

Sarah was following Ellie's lead in search of wildflowers. "Here's a flower I know—clover is attracting bumblebees, too."

"Hmm, we saw sweat bees in the meadow—but maybe more bumblebees up here." Ellie made a note in her journal.

Tony pulled out Henry's map again to locate the second photo point. "Before we hike uphill to take the picture for Henry, would you help me with a monitoring task?" he asked.

"Sure," Ricky answered.

"This'll remind you of the transects we did at the fire site. Walking a straight line from the downhill edge of this grove, to the fence uphill, count any young aspen, no more than four feet tall. I'll mark your starting and ending points on the monitoring map. Here's a data sheet to record your tree counts, and we'll use the map to estimate distances."

Ricky, Ellie, and Sarah each grabbed a data sheet and began walking in almost parallel lines toward the far edge of the grove. On his transect Ricky counted twelve sprouts just a little shorter than he. Ellie had the middle transect, where she had trouble weaving between the trees, almost tripping over a few places where the roots were exposed. She counted five young sprouts. Low branches poked at Sarah, who twisted her way around them; she counted fifteen young trees on her transect. After crouching down to make their way through the trees, they were glad to stand up and stretch before heading uphill.

Tony was taking a photograph for Henry when his helpers joined him. "Thanks, guys. We'll add your records to our long-term data. They'll be tracking this grove for about ten years. Imagine how big the trees will be, how big you'll be, by then!"

The heart-shaped leaves of quaking aspen are attached by long, flattened stems, so the leaves tremble and "quake" in the slightest breeze. Their intermingled roots can sprout quickly after a fire.

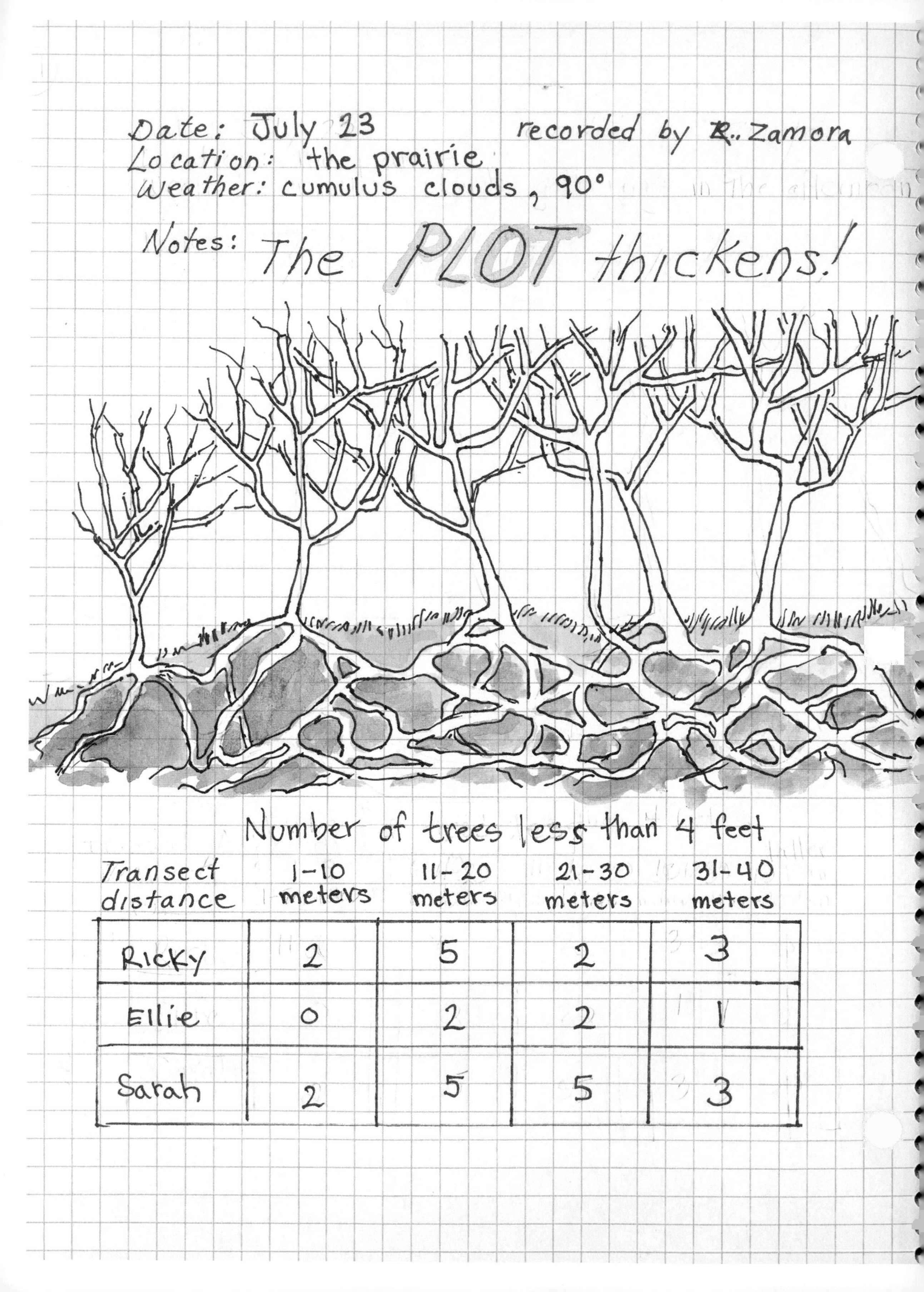

Date: July 23 recorded by R. Zamora
Location: the prairie
Weather: cumulus clouds, 90°

Notes: The PLOT thickens!

Number of trees less than 4 feet

Transect distance	1-10 meters	11-20 meters	21-30 meters	31-40 meters
Ricky	2	5	2	3
Ellie	0	2	2	1
Sarah	2	5	5	3

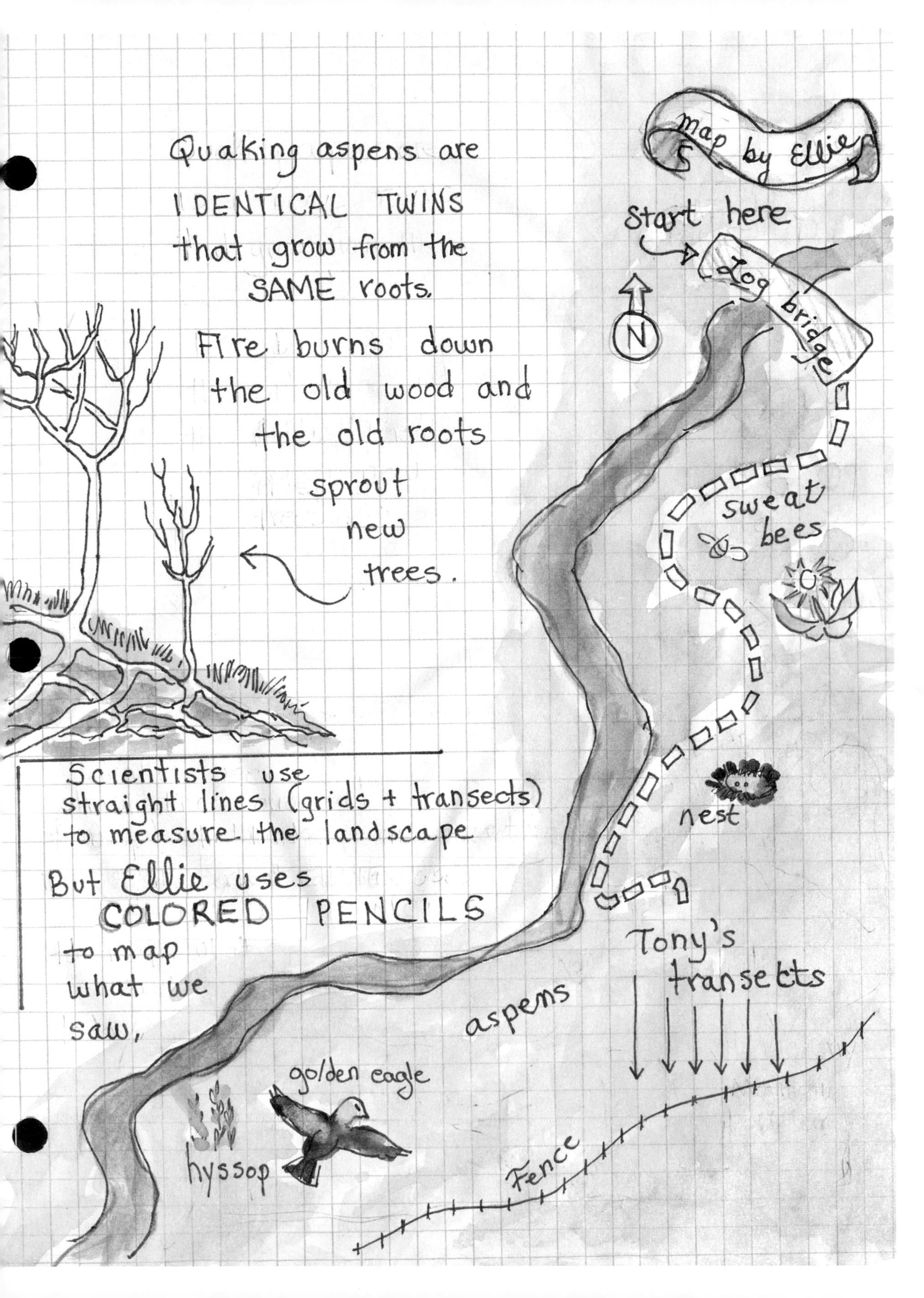

map by Ellie
Quaking aspens are
IDENTICAL TWINS
that grow from the
SAME roots.
Fire burns down
the old wood and
the old roots
sprout
new
trees.
Start here
Log bridge
N
sweat
bees
nest
Scientists use
straight lines (grids + transects)
to measure the landscape
But Ellie uses
COLORED PENCILS
to map
what we
saw,
Tony's
transects
aspens
golden eagle
hyssop
Fence

CHAPTER 8

Time Travellers

The next day they headed over to Henry's house. Sarah knocked on the screen door. "Henry, have I got pictures to show you!"

The old cowboy opened the squeaky door for Ellie, Ricky, and Sarah. "C'mon in. Where's Tony?"

Nighthawks patrol summer evening skies, hunting insects on the wing. Listen for a deep booming sound, like a race car whooshing by, as the male performs his diving, booming display flight.

"Right now he's helping Dad," Sarah replied. "We wanted to come before Ellie goes home."

"Well, glad to see the three of you." Henry opened up his computer so Sarah could pop in her jump drive. "Let's see whatcha got."

"Here's the first photo point looking down on the prairie from the old cottonwood," Sarah said.

Henry compared it to his own photograph. "Look at that! No shrubs in your pic, and I see more flowers, spread out like a wild bouquet. Seems the prairie is changing quickly."

"I loved the flowers," Sarah said. "See my close-ups of the sunny balsamroot that Ellie found, and the lavender lupine?"

"We helped Tony with surveys in the aspen grove before he caught a shot at your second photo point," Ricky explained as the next photo appeared on the screen.

Henry found a copy of his original picture. "What's amazing is that there *is* a grove—see how puny the aspen trees looked before the fencing and burn?"

"Wait a second," Ricky said, signaling to Sarah. "Can you get closer to that bird in the photo?" Perched on the wooden fence was a tannish bird with a black breast band and face mask. "I don't think that's the meadowlark we saw out there," Ricky noted.

"Check out the pair of feathery wisps on his head," Ellie said as Sarah magnified the image. "I think that's a horned lark—not a bird we see at home. Good find, Ricky."

"Hmm. I do believe I'm getting good at field observations." Ricky smiled proudly.

Ellie chuckled. "I'd say it's the company you keep."

Sarah saved a copy of the magnified horned lark and closed down the computer. "Henry, here are the paper copies of your originals," she said.

"No, no—you keep those for your own records. You can be the next photo historian in our valley."

"Thanks a bunch, Henry," she said with a broad grin. "I'll ask Tony to help me, and we'll be a team." Carefully, she tucked the pictures into her backpack. They were ready to head home.

That evening, when the crescent moon was only a slender sliver, a spectacular starry display beckoned the gang out into the yard. "There's so many stars up there, it's hard for me to identify the constellations," Sarah declared.

Ricky remembered visiting cities where lights and smoky air seemed to place a veil across the night skies. "For sure, this super-clear sky gives you a great view," he said. "Look at those

Horned larks can be seen in flocks, poking along bare ground searching for seeds and insects. In recent years their numbers have declined.

three really bright stars that make a summer triangle . . . the small constellation at the bottom of the triangle is Aquila, the eagle.

Sarah followed where Ricky's finger pointed. "Fantastic. An eagle in the night sky!"

"The one I can usually find from there is the big dipper," Ellie added. "And that W-shaped group of stars—that's Cassiopeia. But what blows me away is the great dish-like cloud of stars that is the Milky Way. That's *our* galaxy. We're in the middle of it."

"Billions and billions of stars in our galaxy, and they're almost fourteen billion years old," Tony said.

Ricky shook his head. "I thought the years since dinosaurs were long—but billions of years is a different time scale altogether."

"Y'know, we've been working on maps of places—but now we're talking about mapping out time," Ellie realized.

Ever the mapmaker, Ricky agreed. "Time almost beyond our imagination for stars. Then layers of earth's geologic time starting millions of years ago, and signs left by ancient peoples thousands of years past."

"Trees record changes over hundreds of years," Ellie said, remembering the forests she'd seen coming across the high plateau.

"Henry's photo points, and of course my pictures, record really recent changes," Sarah added.

"Just think—this night sky is so different than that orange sky and fire just a week ago," Ricky reminded them.

"In the bigger scheme of things, recovery from fire might not be all that long," Tony mused. Relaxing together, gazing at the stars, they remembered the aspen, the ponderosa, and the critters they'd seen. Unforgettable images of animals finding refuge high up and down under in the wildfire; plants blooming and sprouting, clear as the night sky.

In the meadow, crickets began their chirping chorus. Then another sound. First, from the hills behind the ranch, some yipping. In response, a coyote howl that was easier to recognize. "Yip, yip, yip" again. "Sounds like someone's learning to howl like a grown-up coyote," Ricky said.

"Magical," Ellie remarked. "A coyote lullaby on my last night here. Thank you, mama coyote. Thanks, everyone, a hot time in a cool place!"

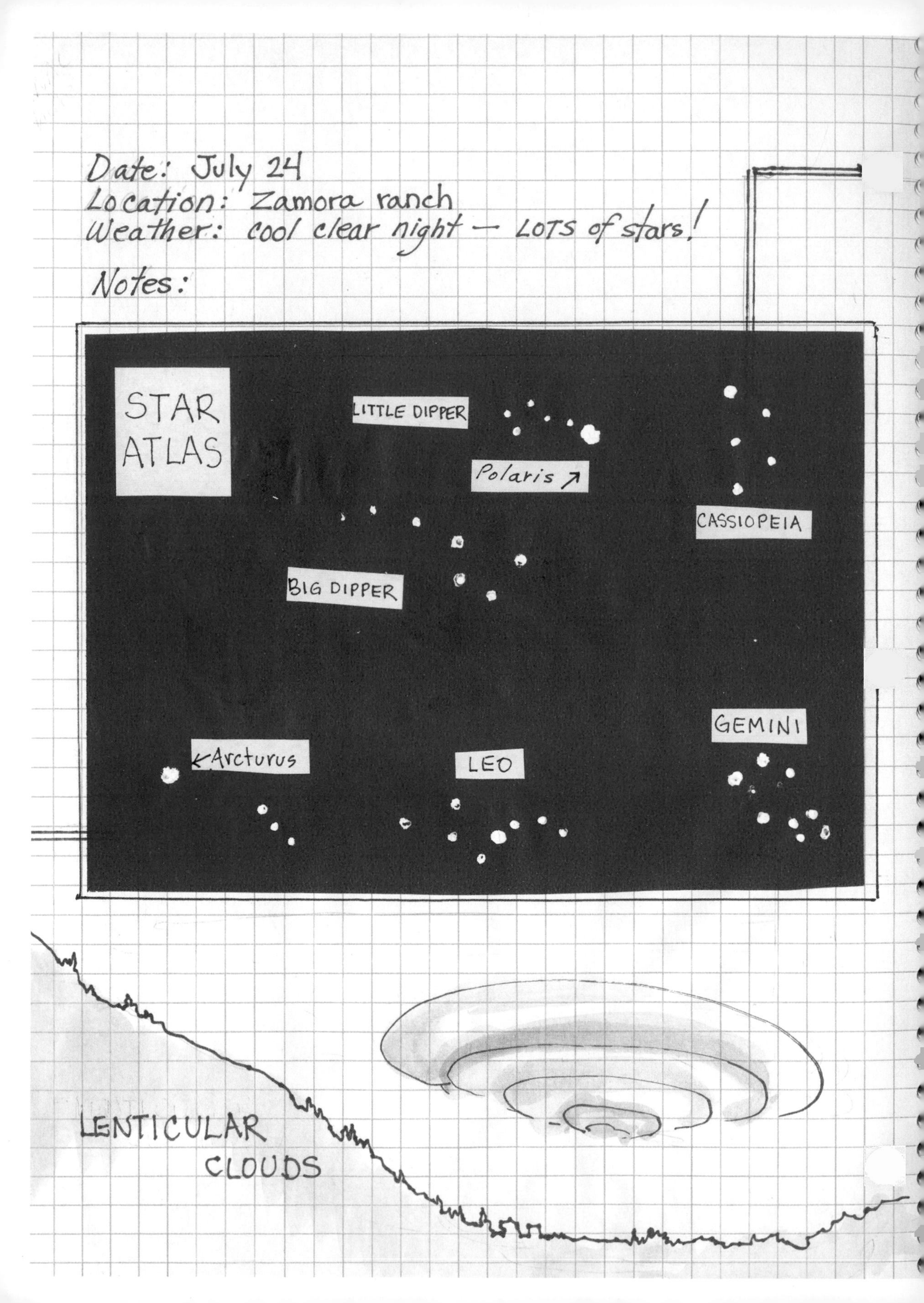
Date: July 24
Location: Zamora ranch
Weather: cool clear night — LOTS of stars!
Notes:
STAR ATLAS
LITTLE DIPPER
Polaris
CASSIOPEIA
BIG DIPPER
GEMINI
Arcturus
LEO
LENTICULAR CLOUDS

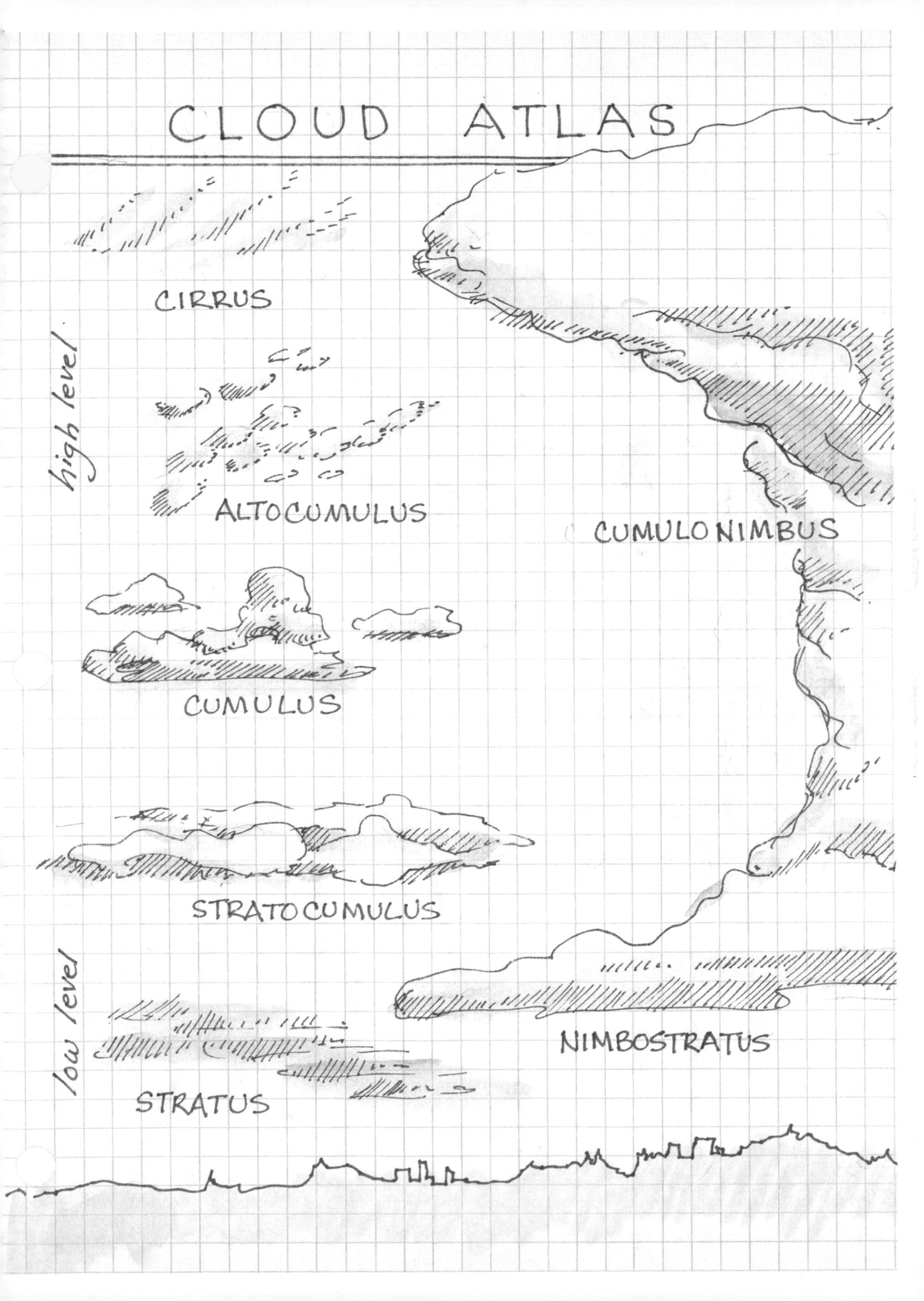
CLOUD ATLAS
CIRRUS
high level
ALTOCUMULUS
CUMULONIMBUS
CUMULUS
STRATOCUMULUS
low level
NIMBOSTRATUS
STRATUS

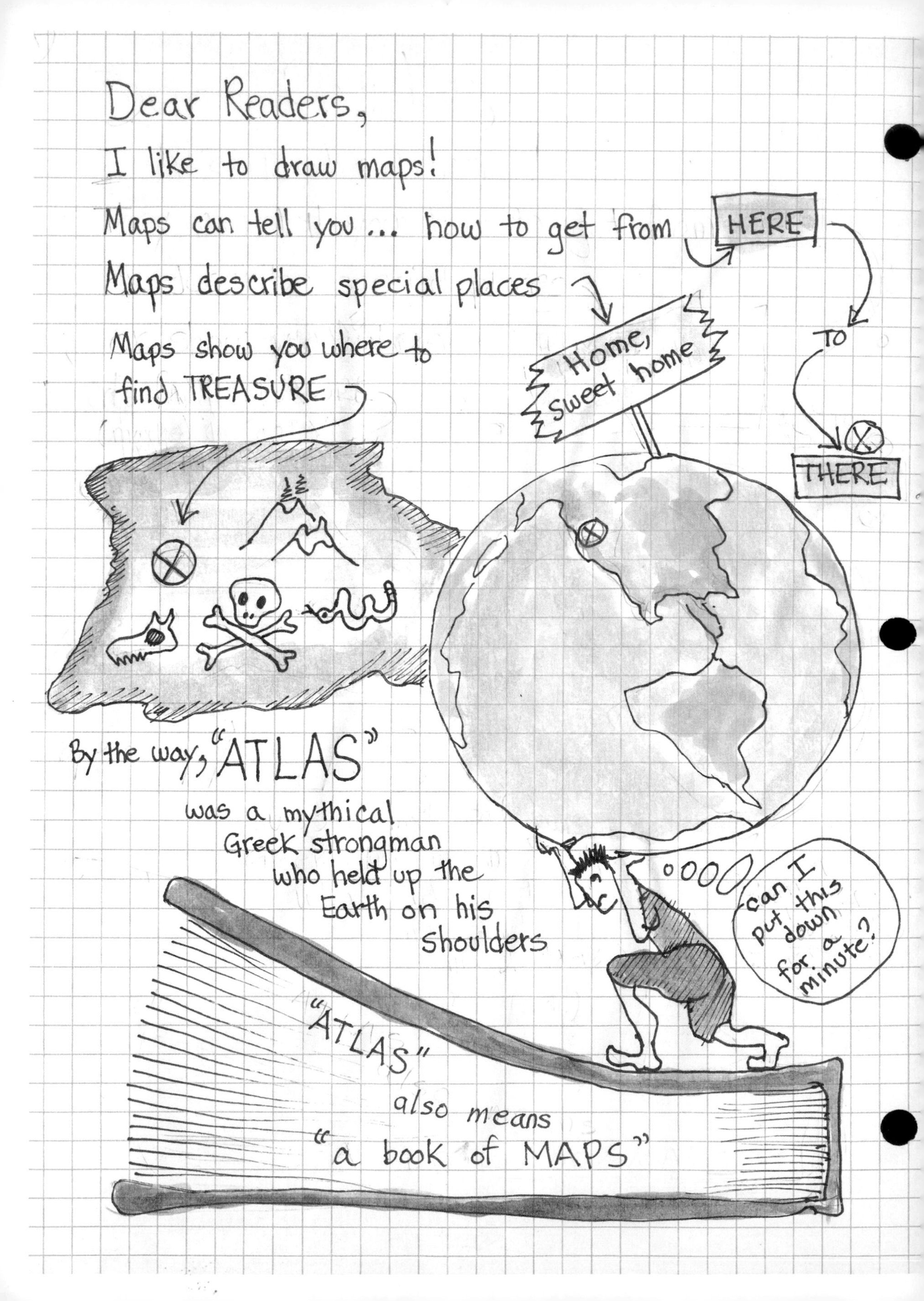
Dear Readers,
I like to draw maps!
Maps can tell you ... how to get from HERE TO THERE
Maps describe special places
Home, sweet home
Maps show you where to find TREASURE
By the way, "ATLAS" was a mythical Greek strongman who held up the Earth on his shoulders
oooO
can I put this down for a minute?
"ATLAS" also means "a book of MAPS"

Some map makers (like my friend Ellie) use COLORS to show features on the map

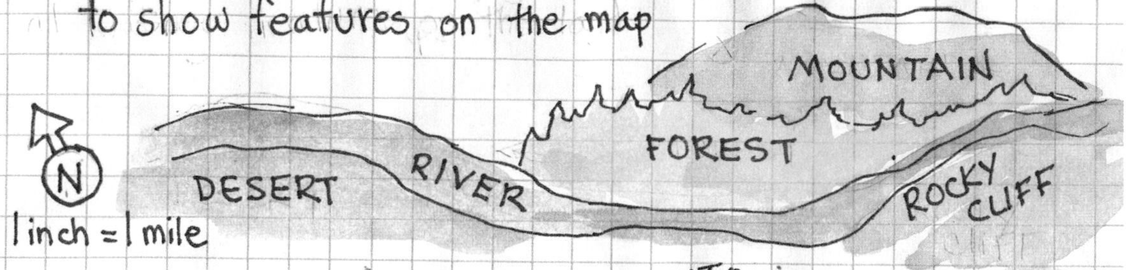

Some map makers (like me) use CONTOUR LINES to show elevation on the map

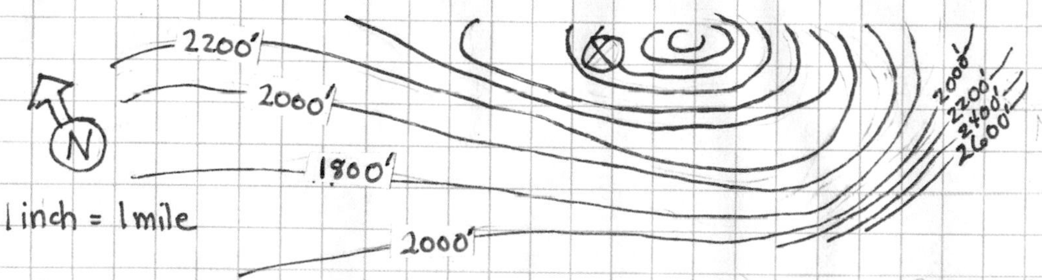

ALL map makers use SCALE...

Also—
I give my maps a DATE: ___
and
a LOCATION: ___
so I will know where I've been and where I'm going.

You can discover a world of TREASURE with maps!

your friend,

Ricky

Ricky's Recommended Reads

About Fire

The Charcoal Forest: How Fire Helps Animals and Plants by Beth Peluso, 2007. Mountain Press. This is a book about recovery from fire in the Northern Rocky Mountains of the United States and Canada. Each page begins with "Who needs burned trees?" then tells about a plant or animal that plays a role in renewal of the forest (there are twenty species). Look for the black-backed woodpecker hiding on each page.

Fire in the Forest: A Cycle of Growth and Renewal by Laurence Pringle, illustrated by Bob Marstall, 1995. Atheneum. Wonderful illustrations follow how the forest recovers after fire. The separate text helped me understand the process of recovery. A good reference book.

About Cartography (Map Making)

Small Worlds: Maps and Mapmaking by Karen Romano Young, 2002. Scholastic Inc. From ancient maps on rocks to views of the universe from space, this book covers the wide world. In the process it showed me how to create maps myself. Maps of our earthly systems, our city networks, even our brains show what we can learn from maps.

Be Your Own Map Expert by Barbara Taylor, 1993. Sterling Publishers. Simple directions for figuring out maps and how to make my own. Liked the ideas for making a compass, fun symbols to use on maps and games, drawing flat and round maps.

Charting the World: Geography and Maps from Cave Paintings to GPS with 21 Activities by Richard Panchyte, 2011. Chicago Review Press. A wonderful book for a map lover like me, filled with old and new maps, besides stories on how they were used. Creative ideas for making all kinds of maps—like engraving a map, projecting a sphere onto a flat surface, making moon observations, and a war strategy game.

About Weather

Weather by Brian Cosgrove, 1991. An Eyewitness Book. Alfred Knopf Inc. Stormy weather of all kinds with helpful labels that explain what is happening in the sky. Kind of an almanac about watching and predicting weather, old records, modern technologies.

About Geology

Focus on Middle School Geology by Rebecca W. Keller, 2012. Gravitas Publications, Inc. Part of the Real Science-4-Kids series. It was easy for me to follow the conversational style of this introduction to geology. Layers inside the Earth, and cycles of the atmosphere and hydrosphere helped me understand what happens on our globe. Plate tectonics, mountain formation, volcano eruptions, and earthquakes are truly amazing.

About Adaptations

Island: A Story of the Galapagos by Jason Chin, 2012. Roaring Brook Press. This book is like a beautifully illustrated diary of an island in little snapshots. From the island's birth six million years ago until 1835 when it sinks back into the ocean, the book records how the island erupted as a volcano, animals that colonized it, new arrivals, and evolutionary changes over time. I like to read this over and over again.

Ecosystem and Field Guides

National Audubon Society First Field Guide Trees by Brian Cassie, 1999. Scholastic Inc. If you'd like to see photographs of the trees in my book, here's where you'll find good shots of cones, needles, leaves, and the trees themselves.

Oregon Wildflowers: A Children's Field Guide to the State's Most Common Flowers by Beverly Magley, illustrated by D. D. Dowden, 1992. Falcon Press I like the simple but realistic color drawings of many of the common flowers I've seen in Oregon. Arrangement by habitats is very handy.

Temperate Grasslands by Ben Hoare, 2011. The Brown Reference Group Ltd. Here's a global view of grasslands that has plants and animals familiar to our native prairie. I especially liked the sections on grassland climate and the ecology of fire in grasslands.

Related Stories

Fire Race: A Karuk Coyote Tale retold by Jonathan London, illustrated by Sylvia Long. 1993. Chronicle Books. The Karuk are native to the Klamath River region of Northern California. This book's vivid illustrations drew me into their traditional tale about how coyote and his friends brought fire from upriver and coaxed it out of willow.

Beaver Steals Fire: A Salish Coyote Story by the Confederated Salish and Kootenai Tribes, illustrated by Sam Sandoval. 2008. Bison Books. This is also a story about how coyote stole fire from the sky world with help from his friends. These tribes presently live in Montana. It would be fun to compare versions. There is also a two-DVD set that goes with this book, *Beaver Steals Fire/Fire on the Land,* which includes the story told by a Salish elder, histories of the two tribes, and information about tribal fire management.

Crow and Weasel by Barry Lopez, illustrated by Tom Pohrt. 1990. Harper Perennial. This story is set in "myth time" by a great storyteller. Though it's not based on any particular Native American tradition, it shows a native respect for the natural world. Crow and Weasel are hunters who travel north through forests and grasslands (set in the northern plains of North America) that remind me of the forests and prairies of eastern Oregon. I hope you like this adventure story about place, survival, and companionship.

While a Tree Was Growing by Jane Bosveld, illustrated by Daniel O'Leary, 1997. American Museum of Natural History. There are two storylines here—one about the giant sequoia that began as a sapling three thousand years ago, and the other about what was happening in human history while the tree was growing. I really got a feel for the time scale of how long old trees have survived on earth.

Websites I Like

http://www.nps.gov/joda/index.htm
Fossils at the John Day Fossil Beds National Monument

http://www.cocorahs.org
This is the Community Collaborative Rain, Hail, and Snow Network page with daily info on precipitation that Tony and I talk about after the big thunderstorm. You can sign up to post what you measure on an official four-inch rain gauge (that you can purchase through them). They have a training video, and the OSU Extension Service has training workshops.

http://oregonseasontracker.forestry.oregonstate.edu/
For our state, Oregon Season Tracker connects information about rainfall and changes in plants during the year.

Acknowledgements

While this is a story with fictional characters, the places they visit and their field experiences are based on real science. The world of *Ricky's Atlas* was created from our research in and journeys to eastern Oregon. These were experiences shared with graduate students, friends, and colleagues. We are grateful to them all.

For expertise shared about particular parts of this book we are indebted to Christine Hirsch, Boone Kaufmann, and Pat Kennedy for their scientific advice. Reviews by Pat Noakes, Bret Hixson (our veterinarian who grew up in eastern Oregon), and two anonymous reviewers were invaluable for improving the authenticity, accuracy, and style of our book. A list of scientific references we used are listed on our website, ellieslog.org. We appreciate the OSU Press Editorial Board, the OSU Libraries, OSU Extension Service 4-H Youth Development Program, and the OSU Department of Fisheries and Wildlife for their support of our project. Many thanks to Faye Chadwell, Donald and Delpha Campbell University Librarian and OSU Press Director, for her extraordinary support of our children's books. We are especially grateful to Mary Braun, Micki Reaman, and Marty Brown at OSU Press for their editorial advice and great care in production and marketing, and to Judy Radovsky, whose dedication in layout design helped us create the book you hold.